# Higher Order Boundary Value Problems on Unbounded Domains

Types of Solutions, Functional Problems and Applications

# TRENDS IN ABSTRACT AND APPLIED ANALYSIS

ISSN: 2424-8746

Series Editor:   John R. Graef
*The University of Tennessee at Chattanooga, USA*

---

*Published*

Vol. 1   Multiple Solutions of Boundary Value Problems:
A Variational Approach
*by John R. Graef & Lingju Kong*

Vol. 2   Nonlinear Interpolation and Boundary Value Problems
*by Paul W. Eloe & Johnny Henderson*

Vol. 3   Solutions of Nonlinear Differential Equations:
Existence Results via the Variational Approach
*by Lin Li & Shu-Zhi Song*

Vol. 4   Quantum Calculus:
New Concepts, Impulsive IVPs and BVPs, Inequalities
*by Bashir Ahmad, Sotiris Ntouyas & Jessada Tariboon*

Vol. 5   Higher Order Boundary Value Problems on Unbounded Domains:
Types of Solutions, Functional Problems and Applications
*by Feliz Manuel Minhós & Hugo Carrasco*

**Trends in Abstract
and Applied Analysis**
Volume **5**

# Higher Order Boundary Value Problems on Unbounded Domains

Types of Solutions, Functional Problems and Applications

Feliz Manuel Minhós

University of Évora, Portugal

Hugo Carrasco

Research Centre in Mathematics and Applications (CIMA), Portugal

**World Scientific**

NEW JERSEY · LONDON · SINGAPORE · BEIJING · SHANGHAI · HONG KONG · TAIPEI · CHENNAI · TOKYO

*Published by*

World Scientific Publishing Co. Pte. Ltd.
5 Toh Tuck Link, Singapore 596224
*USA office:* 27 Warren Street, Suite 401-402, Hackensack, NJ 07601
*UK office:* 57 Shelton Street, Covent Garden, London WC2H 9HE

**Library of Congress Cataloging-in-Publication Data**
Names: Minhós, Feliz Manuel. | Carrasco, Hugo.
Title: Higher order boundary value problems on unbounded domains : types of solutions,
     functional problems and applications / by Feliz Manuel Minhós (University of Évora, Portugal),
     Hugo Carrasco (Research Centre in Mathematics and Applications (CIMA), Portugal).
Description: New Jersey : World Scientific, 2017. |
     Series: Trends in abstract and applied analysis ; volume 5
Identifiers: LCCN 2017015004 | ISBN 9789813209909 (hc : alk. paper)
Subjects: LCSH: Boundary value problems. | Sturm-Liouville equation. | Integral equations.
Classification: LCC QA379 .M55 2017 | DDC 515/.35--dc23
LC record available at https://lccn.loc.gov/2017015004

**British Library Cataloguing-in-Publication Data**
A catalogue record for this book is available from the British Library.

Desk Editors: V. Vishnu Mohan/Kwong Lai Fun

Typeset by Stallion Press
Email: enquiries@stallionpress.com

Printed in Singapore

# Contents

*Preface*                                                                   ix

*Introduction*                                                              xi

## Part I. Boundary Value Problems on the Half-Line    1

### Introduction                                                            3

1.  Third-Order Boundary Value Problems                                     5

    1.1   Introduction . . . . . . . . . . . . . . . . . . . . . . . . . . . . .    5
    1.2   Definitions and auxiliary results . . . . . . . . . . . . . .    6
    1.3   Existence and localization result . . . . . . . . . . . . .    11
    1.4   Example . . . . . . . . . . . . . . . . . . . . . . . . . . . . . .    17

2.  General $n$th-Order Problems                                            19

    2.1   Introduction . . . . . . . . . . . . . . . . . . . . . . . . . . . . .    19
    2.2   Preliminary results . . . . . . . . . . . . . . . . . . . . . . .    20
    2.3   Existence and localization result . . . . . . . . . . . . .    22
    2.4   Example . . . . . . . . . . . . . . . . . . . . . . . . . . . . . .    23

3.  Impulsive Problems on the Half-Line with Infinite
    Impulse Moments                                                   25

    3.1   Introduction . . . . . . . . . . . . . . . . . . . . . . . . . . . . .    25
    3.2   Definitions and preliminary results . . . . . . . . . . . .    26

3.3   Main result . . . . . . . . . . . . . . . . . . . . . . . .   28

3.4   Example . . . . . . . . . . . . . . . . . . . . . . . . . .   36

# Part II. Homoclinic Solutions and Lidstone Problems     39

## Introduction     41

4.   Homoclinic Solutions for Second-Order Problems     43

4.1   Introduction . . . . . . . . . . . . . . . . . . . . . . . .   43

4.2   Preliminaries . . . . . . . . . . . . . . . . . . . . . . . .   44

4.3   Existence and localization of homoclinics . . . . . . . . . .   47

4.4   Example of a discontinuous BVP . . . . . . . . . . . . . .   52

4.5   Duffing equation . . . . . . . . . . . . . . . . . . . . . .   53

4.6   Forced cantilever beam equation with damping . . . . . .   54

5.   Homoclinic Solutions to Fourth-Order Problems     57

5.1   Introduction . . . . . . . . . . . . . . . . . . . . . . . .   57

5.2   Definitions and auxiliary results . . . . . . . . . . . . . .   59

5.3   Existence results . . . . . . . . . . . . . . . . . . . . . .   61

5.4   Example . . . . . . . . . . . . . . . . . . . . . . . . . . .   65

5.5   Bernoulli–Euler–v. Karman problem . . . . . . . . . . . .   66

5.6   Extended Fisher–Kolmogorov and
      Swift–Hohenberg problems . . . . . . . . . . . . . . . . . .   67

6.   Lidstone Boundary Value Problems     71

6.1   Introduction . . . . . . . . . . . . . . . . . . . . . . . .   71

6.2   Auxiliary definitions and Green's functions . . . . . . . .   73

6.3   Existence result . . . . . . . . . . . . . . . . . . . . . .   75

6.4   An infinite beam resting on granular foundations . . . . .   78

# Part III. Heteroclinic Solutions and Hammerstein Equations     81

## Introduction     83

7.   Heteroclinic Solutions for Semi-linear Problems (i)     85

7.1   Introduction . . . . . . . . . . . . . . . . . . . . . . . .   85

7.2 Definitions and preliminary results . . . . . . . . . . . . 87
7.3 Existence of heteroclinics . . . . . . . . . . . . . . . . . 89
7.4 Example . . . . . . . . . . . . . . . . . . . . . . . . . . . . 94

8. Heteroclinic Solutions for Semi-linear Problems (ii) 97
   8.1 Introduction . . . . . . . . . . . . . . . . . . . . . . . . . 97
   8.2 Auxiliary results . . . . . . . . . . . . . . . . . . . . . . . 99
   8.3 Existence of heteroclinics solutions . . . . . . . . . . . . 102
   8.4 Examples . . . . . . . . . . . . . . . . . . . . . . . . . . . 108

9. Heteroclinic Solutions for Semi-linear Problems (iii) 111
   9.1 Introduction . . . . . . . . . . . . . . . . . . . . . . . . . 111
   9.2 Existence results . . . . . . . . . . . . . . . . . . . . . . . 113
   9.3 Example . . . . . . . . . . . . . . . . . . . . . . . . . . . . 124
   9.4 Singular $\phi$-Laplacian equations . . . . . . . . . . . . . . 125

10. Hammerstein Integral Equations with Sign-Changing Kernels 127
    10.1 Introduction . . . . . . . . . . . . . . . . . . . . . . . . . 127
    10.2 Main result . . . . . . . . . . . . . . . . . . . . . . . . . . 129
    10.3 Application to fourth-order BVPs and infinite beams . . . 131

**Part IV. Functional Boundary Value Problems** **133**

**Introduction** 135

11. Second-Order Functional Problems 139
    11.1 Introduction . . . . . . . . . . . . . . . . . . . . . . . . . 139
    11.2 Definitions and auxiliary results . . . . . . . . . . . . . . 140
    11.3 Existence and localization results . . . . . . . . . . . . . 142
    11.4 Example . . . . . . . . . . . . . . . . . . . . . . . . . . . . 149
    11.5 Emden–Fowler equation . . . . . . . . . . . . . . . . . . . 149

12. Third-Order Functional Problems 153
    12.1 Introduction . . . . . . . . . . . . . . . . . . . . . . . . . 153
    12.2 Definitions and *a priori* bounds . . . . . . . . . . . . . . 154
    12.3 Existence and localization results . . . . . . . . . . . . . 159
    12.4 Falkner–Skan equation . . . . . . . . . . . . . . . . . . . . 168

13. $\phi$-Laplacian Equations with Functional Boundary Conditions        171

    13.1  Introduction . . . . . . . . . . . . . . . . . . . . . . . . 171
    13.2  Preliminary results . . . . . . . . . . . . . . . . . . . . 173
    13.3  Existence and localization result . . . . . . . . . . . . 181
    13.4  Examples . . . . . . . . . . . . . . . . . . . . . . . . . 186

*Bibliography*        189

*Index*        199

# Preface

The relative scarcity of results that guarantee the existence of solutions for boundary value problems on unbounded domains, contrasts with the high applicability on real problems with differential equations defined on the half-line or on the whole real line. This gap is the main reason that led to this work.

This book contains four parts with different problems composed by differential equations, from second to higher orders, and integral Hammerstein equations, several types of boundary conditions, for example, Sturm–Liouville, Lidstone and functional conditions, and solutions with diverse qualitative properties, such as impulsive, homoclinic, and hetero-clinic solutions.

The noncompactness of the time interval and the possibility of studying the unbounded functions will require the definition of adequate Banach spaces. In fact, the space considered, the functional framework assumed and the set of admissible solutions for each problem are defined under a main goal: the functions must remain bounded for the space and the norm in consideration. This is achieved by defining some weight functions (polynomial or exponential) in the space or assuming some asymptotic behavior.

We underline some new features of the content:

- relation between some properties of the Green's functions defined on the real line, the existence of homoclinic solutions and the solvability of Lidstone-type problems;
- existence of heteroclinic solutions for semi-linear problems without growth or asymptotic assumptions on the nonlinearity;

- solvability of Hammerstein integral equations on the whole line, with discontinuous and sign-changing kernels and with nonlinear dependence on several derivatives.

In addition to the existence, solutions will be localized in a strip. The lower and upper solutions method will play an important role, and combined with other tools like the one-sided Nagumo growth conditions, Green's functions or Schauder's fixed-point theorem, provide the existence and location results for differential equations with various boundary conditions.

Different applications to real phenomena will be presented, most of them translated into classical equations as Duffing, Bernoulli–Euler–v. Karman, Fisher–Kolmogorov, Swift–Hohenberg, Emden–Fowler or Falkner–Skan-type equations.

All these applications have a common denominator: they are defined in unbounded intervals and the existing results in the literature are scarce or proven only numerically in discrete problems.

*Feliz Manuel Minhós*
*Hugo Carrasco*

# Introduction

The *leitmotiv* of this book is related with higher order boundary value problems (BVPs) defined on unbounded domains, more precisely on the half-line or on the whole real line.

Roughly speaking, we can say that BVPs are rather different from initial (or final) value problems as they do not have a continuous dependence on the boundary data. In fact, small perturbations on boundary values may cause vital changes on the qualitative properties of the corresponding solutions, and even on the existence, nonexistence or multiplicity of solutions. The following example will illustrate this fact.

Consider the second-order homogeneous differential equation

$$y'' + y = 0. \tag{1}$$

The initial value problem, known as Cauchy problem, composed by (1) and the initial values

$$y(0) = k_1, \quad y'(0) = k_2$$

has a unique solution given by $y(x) = k_1 \cos x + k_2 \sin x$, for every real $k_1, k_2$.

However, the BVP with (1) and the Dirichlet boundary conditions

$$y(0) = 0, \quad y(\pi) = \varepsilon (\neq 0)$$

has no solution, but the Dirichlet BVP with (1) and

$$y(0) = 0, \quad y(\beta) = \varepsilon, \quad \text{with } 0 < \beta < \pi,$$

has a unique solution, $y(x) = \frac{\varepsilon \sin x}{\sin \beta}$, and the BVP composed by (1) together with the boundary conditions

$$y(0) = 0, \quad y(\pi) = 0,$$

has infinite solutions of the type $y(x) = c \sin x$, with arbitrary $c \in \mathbb{R}$.

In the past decades, the study of BVPs defined on compact intervals has been considered by many authors with application of a huge variety of methods and techniques. However, BVPs defined on unbounded intervals are scarce, as they require other types of techniques to overcome the lack of compactness.

Historically, these problems began at the end of nineteenth century with A. Kneser. This pioneer work described monotone solutions of second-order ordinary differential equations. Others followed his results and different techniques have been studied, namely the lower and upper solutions method (see [13] and the references therein).

Several real problems were modeled by BVPs defined on infinite intervals. As examples, we refer to the study of unsteady flow of a gas through a semi-infinite porous medium; the discussion of electrostatic probe measurements in solid-propellant rocket exhausts; the analysis of the mass transfer on a rotating disk in a non-Newtonian fluid; the heat transfer in the radial flow between parallel circular disks; the investigation of the temperature distribution in the problem of phase change of solids with temperature-dependent thermal conductivity, as well as numerous problems arising in the study of draining flows, circular membranes, plasma physics, radially symmetric solutions of semi-linear elliptic equations, nonlinear mechanics, and non-Newtonian fluid flows; and the bending of infinite beams and its applications in the railways and highways. More details and examples can be seen in [5] and the references therein.

This book is divided into four parts, each one related to some type of BVPs on unbounded intervals.

The first part, *Boundary Value Problems on the Half-Line*, is dedicated to higher order BVPs, defined on the half-line, and it is composed of three chapters:

- Chapter 1 — *Third-Order Boundary Value Problems*. Third-order differential equations on infinite intervals can describe the evolution of physical phenomena like draining or coating fluid flow problems. The noncompactness of the time interval and the possibility of studying unbounded functions require the redefinition of the admissible Banach space and its weighted norms. This chapter will prove the existence and localization of, at least, one solution for a BVP with Sturm–Liouville-type boundary conditions. The tools involved will be the one-sided Nagumo-type growth condition, Green's functions, lower and upper solutions method and Schauder's fixed-point theorem. An example will conclude the chapter.

- Chapter 2 — *General nth-Order Problems.* This chapter arises from an attempt to generalize the previous one to order $n$. In a particular case, fourth-order differential equations can model the bending of an elastic beam. An example is shown to demonstrate the importance of the one-sided Nagumo-type growth condition.
- Chapter 3 — *Impulsive Problems on the Half-Line with Infinite Impulse Moments.* Some of the previous techniques are applied in a second-order impulsive problem on the half-line, with generalized impulsive functions, depending on the unknown function and its derivative, and allowing an infinite number of impulse moments. The notion of Carathéodory sequence is a key argument in the method.

The second part, *Homoclinic Solutions and Lidstone Problems,* considers BVPs on the whole real line, looking for sufficient conditions on the nonlinearity to guarantee the existence of homoclinic solutions, and its relation to solutions for Lidstone-type problems. It contains three chapters:

- Chapter 4 — *Homoclinic Solutions for Second-Order Problems.* In this chapter, the lower and upper solutions method will be used with unordered functions. An existence and localization result will be settled. Specific applications to Duffing-type equations and beam equations with damping will conclude the chapter.
- Chapter 5 — *Homoclinic Solutions to Fourth-Order Problems.* Different problems involving Bernoulli–Euler–v. Karman, Fisher–Kolmogorov or Swift–Hohenberg equations are strongly linked with fourth-order differential equations. This chapter will establish the existence results and examples for each particular case.
- Chapter 6 — *Lidstone Boundary Value Problems.* The Lidstone theory, initially applied to interpolation problems, is considered, in this chapter, in the whole real line with a strong connection to the homoclinic solutions. In this final chapter of this part, a problem of an infinite beam resting on granular foundations with moving loads will be studied.

The third part, *Heteroclinic Solutions and Hammerstein Equations,* contains four chapters:

- Chapters 7–9 provide sufficient conditions for the existence of heteroclinic solutions for three types of $\phi$-Laplacian equations, sometimes named as semi-linear equations, on the real line. We point out that these heteroclinic solutions are obtained without the usual monotone or growth assumptions on the nonlinearity.

- Chapter 10 studies integral equations, more precisely, Hammerstein equations, defined on the whole real line, with discontinuous nonlinearities, which may depend, not only on the unknown function, but also on some derivatives, without monotone or asymptotic assumptions. Moreover, the kernels and their partial derivatives in order to the first variable, are very general functions: they may be discontinuous and may change the signal. A simple criterion is included to see if the existing solutions are homoclinic or heteroclinic solutions, together with an application to a fourth-order BVP.

In the last part, *Functional Boundary Value Problems*, we study BVPs with functional boundary conditions, that is, with boundary data that can depend globally on the correspondent variables. In this way, it contains and generalizes many types of boundary conditions such as multipoint, advanced or delayed, nonlocal, integro-differential, with maximum or minimum arguments, among others. Part IV is divided into three chapters, each one with a different type of problems:

- Chapter 11 — *Second-Order Functional Problems.* BVPs involving functional boundary conditions can model thermal conduction, semiconductor and hydrodynamic problems. An application to a problem composed by an Emden–Fowler-type equation and an infinite multipoint condition will be formulated and solved.
- Chapter 12 — *Third-Order Functional Problems.* Falkner–Skan equations are obtained from partial differential equations. They can model the behavior of a viscous flow over a plate. Until now, only numerical techniques could deal with this type of problems, however, this chapter will prove an existence and localization result by topological methods.
- Chapter 13 — *Phi-Laplacian Equations with Functional Boundary Conditions.* This final chapter will deal with weighted norms, namely the Bielecki norm. This will be a fundamental tool to manage unbounded solutions. An important fact is that the homeomorphism $\phi$ does not need to be surjective.

Throughout this work, the usual lemma of Arzèla–Ascoli could not be used due to lack of compactness, and this issue is overcome with some methods, techniques and specific tools. We point out some of them:

- **Weighted spaces** and the corresponding **weighted norms**;
- **Carathéodory functions** admissible for the nonlinearities;

- **Green's functions** on unbounded domains;
- **Equiconvergence** at $\infty$.

The space considered and the functional framework assumed define the set of admissible solutions for each problem with a main goal: the functions must remain bounded for the space and the norm in consideration. This is achieved by defining some weight functions (polynomial or exponential) in the space or assuming some asymptotic behavior. Therefore, for each problem, the specific space and norm to be used are presented.

The type of nonlinearities in the different problems has a common feature: roughly, they must be measurable in the time variable, continuous almost everywhere, on the space variables, and having a growth controlled by an $L^1$-function on $[0, +\infty[$ or $\mathbb{R}$. A function with such properties is called in the literature as an $L^1$-Carathéodory function. To avoid boring repetitions we define them for an general unbounded interval $I$ (see Definition 1.2.1), which will be the half-line, or the whole real line, according to each problem.

The Green's functions and their properties play a key role in some problems, for which we carry out more detailed considerations.

Basically, these functions are solutions of a linear BVP, irrespective of whether they are homogeneous or not, and they will guarantee the existence of at least one solution, and, moreover, they can provide the explicit expression of the solution for the studied BVP. In a broader sense, they can be seen as a particular case of the so-called kernel functions, as they are related with the kernel of linear operators.

When dealing with linear and homogeneous ordinary differential equations on the form

$$Lu(t) = 0,$$

it is clear that any homogeneous solution is a linear combination of some independent functions (in the same number as the degree of the ODE). However, when the differential equation is nonhomogeneous

$$Lu(t) = e(t), \tag{2}$$

it is fundamental to find a particular solution for each function $e$ and then add it to the linear combination referred.

The Green's functions method is due to George Green (1793–1841), the first mathematician to use such kind of kernels to solve BVPs.

If equation (2), coupled with homogeneous boundary conditions, has only the trivial solution for $e(t) = 0$, then the associated linear operator is invertible and its inverse operator, $L^{-1}e$, is characterized with an integral kernel, $G(t, s)$, called the Green's function. The solution of this problem is then given by

$$u(t) = L^{-1}e(t) := \int_a^b G(t, s)e(s)ds, \quad \forall t \in [a, b].$$

A remarkable characteristic of the explicit expression of the Green's functions is the fact that it is independent on the function $e$. After that, one needs to calculate the integral expression and then it is possible to obtain some additional qualitative information about solutions: sign, oscillation properties, *a priori* bounds or their stability. All these issues transform the theory of Green's functions in a fundamental tool in the analysis of differential equations. It has been widely studied in the literature and reveals to be very important in order to use monotone iterative techniques, lower and upper solutions, fixed point theorems or variational methods (see [39] and references therein).

The equiconvergence at $\infty$, sometimes called as the stability at $\infty$, is a crucial argument to recover the compacity of the operator on unbounded domains. Indeed, with such concept, we can formulate a criterion that plays the role of the Arzèla–Ascoli theorem for bounded domains. More precisely, if, in some subset $M$ of the space, the functions are uniformly bounded, equicontinuous on some subintervals of $[0, \infty)$ or $\mathbb{R}$, and equiconvergent at $\infty$, or $\pm\infty$, then $M$ is relatively compact.

As it can easily be seen, the above notion depends on the space considered, the weights defined, and on the order of the derivatives involved. Therefore, for the reader's convenience, we specify in each problem the detailed criterion referred.

Finally, we point out that in all chapters there are examples to illustrate each theorem or, even, concrete applications to real phenomena.

Part I

# Boundary Value Problems
# on the Half-Line

# Introduction

Sturm–Liouville theory was initiated by Jacques Charles François Sturm (1803–1855) and Joseph Liouville (1809–1882) to study second-order linear differential equations of the form

$$\frac{d}{dt}\left(p(t)\frac{dy}{dt}\right) + (\lambda w(t) - q(t))y = 0,$$

where $p, q$ are positive functions, $\lambda$ is a constant and $w$ is a known function called either the density or weighting function.

The common approach to this equation deals with bounded intervals, that is, $t \in [a, b]$, $a, b \in \mathbb{R}$, $a < b$, and with boundary conditions of the form

$$c_1 y(a) + c_2 y'(a) = 0, \quad c_3 y(b) + c_4 y'(b) = 0, \quad c_1, c_2, c_3, c_4 \in \mathbb{R}.$$

This kind of boundary conditions will, in this first part, be generalized to third and $n$th-order BVPs, defined on unbounded intervals. Thus, in what follows, BVPs with Sturm–Liouville boundary conditions may also be called simply as Sturm–Liouville problems.

The great novelty of this part is to assume a one-sided Nagumo condition. In fact, the usual bilateral Nagumo condition used in the literature requires a subquadratic growth for the nonlinearities. As far as we know, it is the first time where the unilateral Nagumo conditions are adapted to unbounded domains. In this way, the nonlinearities may have an asymmetric growth, being, for example, asymptotically unbounded on one side and retaining the subquadratic growth on the other side.

This first part is separated into three chapters, dealing with problems defined on the half-line.

In Chapter 1, the existence of at least one solution for a BVP involving a third-order differential equation is proved, and it is based on [117]. Other

3

properties are proved for such solutions like localization and asymptotic properties.

Chapter 2 is assigned to a generic $n$th-order problem, where the main result is an existence and localization result, meaning that it provides not only the existence but also the localization of the unknown function and its derivatives via lower and upper solutions method.

In Chapter 3, the previous techniques are applied to a second-order impulsive problem in the half-line with a full nonlinearity and infinite impulsive effects, on the unknown function and its first derivative, given by generalized functions. The notion of Carathéodory sequences and the equi-convergence at $+\infty$ and at each impulsive moment are key arguments to have a compact operator.

Lower and upper solutions method is a useful technique to deal with BVPs as, from their localization part, some qualitative data about solution variation and behavior can be obtained (see [32, 71, 99, 113, 120]). Another important tool is the Nagumo condition, useful to obtain *a priori* estimates on some derivative of the solution, generalizing subquadratic growth assumptions on the nonlinear part of the differential equation.

As it can be seen in the references above, the usual growth condition of the Nagumo type is a bilateral one. However, the same estimation holds with a similar one-sided assumption, allowing that the BVPs can include unbounded nonlinearities. In this way, it generalizes the two-sided condition, as it is proved in [62, 75].

Finally, it is worth mentioning that, in both chapters, the nonlinearities are $L^1$-Carathéodory functions and, therefore, they may have discontinuities in time.

# Chapter 1

# Third-Order Boundary Value Problems

## 1.1. Introduction

Third-order differential equations arise in many areas, such as the deflection of an elastic beam having a constant or varying cross-section, three-layer beam, electromagnetic waves or gravity-driven flows (see [73] and the references therein).

In infinite intervals, third-order BVPs can describe the evolution of physical phenomena, for example, some draining or coating fluid-flow problems (see [139]).

Due to the noncompactness of the interval, the discussion about sufficient conditions for the solvability of BVPs is more delicate. In the literature, existence results to such problems are, mainly, due to the extension of continuous solutions on the corresponding finite intervals, under a diagonalization process and fixed point theorems, in special Banach spaces (see [4, 19, 98, 146] and the references therein).

The present chapter will study a general Sturm–Liouville-type BVP, composed by a third-order differential equation defined on the half-line

$$u'''(t) = f(t, u(t), u'(t), u''(t)), \quad \text{a.e. } t \geq 0 \tag{1.1.1}$$

together with boundary conditions

$$u(0) = A, \quad au'(0) + bu''(0) = B, \quad u''(+\infty) = C, \tag{1.1.2}$$

with $f : \mathbb{R}_0^+ \times \mathbb{R}^3 \to \mathbb{R}$ an $L^1$-Carathéodory function (eventually discontinuous on time), where $u''(+\infty) := \lim_{t \to +\infty} u''(t)$, $a, b, A$, $B, C \in \mathbb{R}$ and $a > 0, b < 0$.

The setback of dealing with unbounded intervals and the possibility of studying unbounded functions can be overcome with new definitions of weighted spaces and norms.

## 1.2. Definitions and auxiliary results

As solutions can be unbounded, the functional framework must be defined with some weight functions and the corresponding weighted norms.

Consider the space

$$X_1 = \left\{ x \in C^2(\mathbb{R}_0^+) : \lim_{t \to +\infty} \frac{x^{(i)}(t)}{\omega_i(t)} \in \mathbb{R}, \ i = 0, 1, 2 \right\}$$

with $\omega_i(t) = 1 + t^{2-i}, i = 0, 1, 2$ and the norm

$$\|x\|_{X_1} = \max \left\{ \|x\|_0, \|x'\|_1, \|x''\|_2 \right\},$$

where

$$\|y\|_i = \sup_{t \geq 0} \left| \frac{y(t)}{\omega_i(t)} \right|, \quad \text{for } i = 0, 1, 2.$$

By standard arguments, it can be proved that $(X_1, \|\cdot\|_{X_1})$ is a Banach space.

Let us express the concept of $L^1$-Carathéodory functions to be used forward.

**Definition 1.2.1.** Let $E$ be a normed space and $I$ be an unbounded interval $(I = \mathbb{R}_0^+ \text{ or } I = \mathbb{R})$.

A function $f : I \times \mathbb{R}^n \to \mathbb{R}$ is $L^1$-Carathéodory if it verifies the following conditions:

(i) for each $\xi \in \mathbb{R}^n$, $t \mapsto f(t, \xi)$ is measurable on $I$;
(ii) for almost every $t \in I$, $\xi \mapsto f(t, \xi)$ is continuous in $\mathbb{R}^n$;
(iii) for each $\rho > 0$, there exists a positive function $\varphi_\rho \in L^1(I)$ such that, for $\|\xi\|_E < \rho$,

$$|f(t, \xi)| \leq \varphi_\rho(t), \quad \text{a.e. } t \in I.$$

For each particular structure of the space $E$, and the corresponding norm, condition (iii) assumes different forms of inequalities.

Let $\gamma_i, \Gamma_i \in C(\mathbb{R}_0^+)$, such that $\gamma_i(t) \le \Gamma_i(t), \forall t \ge 0, i = 0, 1$ and

$$E_1 = \left\{ (t, x_0, x_1, x_2) \in \mathbb{R}_0^+ \times \mathbb{R}^3 : \gamma_i(t) \le x_i \le \Gamma_i(t), i = 0, 1 \right\}.$$

The following one-sided Nagumo condition generalizes the usual bilateral one.

**Definition 1.2.2.** A function $f : E_1 \to \mathbb{R}$ is said to satisfy a one-sided Nagumo-type growth condition in $E_1$ if, for some positive and continuous functions $\psi, h$ and some $\nu > 1$, such that

$$\int_0^{+\infty} \psi(s)ds < +\infty, \quad \sup_{t \ge 0} \psi(t)(1+t)^\nu < +\infty, \quad \int_0^{+\infty} \frac{s}{h(s)}ds = +\infty,$$

$$(1.2.1)$$

it verifies either

$$f(t, x, y, z) \le \psi(t)h(\|z\|_2), \quad \forall(t, x, y, z) \in E_1 \qquad (1.2.2)$$

or

$$f(t, x, y, z) \ge -\psi(t)h(\|z\|_2), \quad \forall(t, x, y, z) \in E_1. \qquad (1.2.3)$$

An important goal of this condition is to give an *a priori* bound on the second derivative of all existent solutions.

**Lemma 1.2.3.** *Let $f : \mathbb{R}_0^+ \times \mathbb{R}^3 \to \mathbb{R}$ be an $L^1$-Carathéodory function satisfying (1.2.1) and, either (1.2.2) or (1.2.3) in $E_1$. Then there exists $R > 0$ (not depending on $u$) such that every solution $u$ of (1.1.1),(1.1.2) satisfying*

$$\gamma(t) \le u(t) \le \Gamma(t), \gamma'(t) \le u'(t) \le \Gamma'(t), \quad \forall t \ge 0 \qquad (1.2.4)$$

*verifies $\|u''\|_2 < R$.*

**Proof.** Let $u$ be a solution of (1.1.1), (1.1.2) verifying (1.2.4). Consider $r > 0$ such that

$$r > \max \left\{ \left| \frac{B - a\Gamma'(0)}{b} \right|, \left| \frac{B - a\gamma'(0)}{b} \right|, |C| \right\}. \qquad (1.2.5)$$

By the previous inequality, it is impossible that $|u''(t)| > r, \forall t \geq 0$ because

$$|u''(0)| = \left|\frac{B - au'(0)}{b}\right| \leq \max\left\{\left|\frac{B - a\Gamma'(0)}{b}\right|, \left|\frac{B - a\gamma'(0)}{b}\right|\right\} < r.$$

If $|u''(t)| \leq r, \forall t \geq 0$, taking $R > \frac{r}{2}$, the proof is complete as

$$\|u''\|_2 = \sup_{t \geq 0}\left|\frac{u''(t)}{2}\right| \leq \frac{r}{2} < R.$$

In the following, it will be proved that even when there exists $t > 0$ such that $|u''(t)| > r$, the norm $\|u''\|_2$ remains bounded, in all possible cases, $f$ verifies either (9.2.13) or (1.2.3).

Suppose there exists $t > 0$ such that $|u''(t)| > r$, that is, $u''(t) > r$ or $u''(t) < -r$. In the first case, by (1.2.1), one can take $R > r$ such that

$$\int_r^R \frac{s}{h(s)}ds > M \max\left\{M_1 + \sup_{t \geq 0}\frac{\Gamma'(t)}{1+t}\frac{\nu}{\nu - 1}, M_1 - \inf_{t \geq 0}\frac{\gamma'(t)}{1+t}\frac{\nu}{\nu - 1}\right\}$$

with $M := \sup_{t \geq 0}\psi(t)(1 + t)^\nu$ and $M_1 := \sup_{t \geq 0}\frac{\Gamma'(t)}{(1+t)^\nu} - \inf_{t \geq 0}\frac{\gamma'(t)}{(1+t)^\nu}$.

If condition (1.2.2) holds, then, by (1.2.5), there are $t_*, t_+ \in \mathbb{R}^+$ such that $t_* < t_+, u''(t_*) = r$ and $u''(t) > r, \forall t \in (t_*, t_+]$. Therefore,

$$\int_{u''(t_*)}^{u''(t_+)} \frac{s}{h(s)}ds = \int_{t_*}^{t_+} \frac{u''(s)}{h(u''(s))}u'''(s)ds \leq \int_{t_*}^{t_+} \psi(s)u''(s)ds$$

$$\leq M \int_{t_*}^{t_+} \frac{u''(s)}{(1+s)^\nu}ds$$

$$= M \int_{t_*}^{t_+} \left[\left(\frac{u'(s)}{(1+s)^\nu}\right)' + \frac{\nu u'(s)}{(1+s)^{1+\nu}}\right] ds$$

$$\leq M \left(M_1 + \sup_{t \geq 0}\frac{\Gamma'(t)}{1+t}\int_0^{+\infty} \frac{\nu}{(1+s)^\nu}ds\right) < \int_r^R \frac{s}{h(s)}ds.$$

Consequently, $u''(t_+) < R$ and as $t_*$ and $t_+$ are arbitrary in $\mathbb{R}^+$, then $u''(t) < R, \forall t > 0$. Similarly, the case where there are $t_-, t_* \in \mathbb{R}^+$ can be proved such that $t_- < t_*$ and $u''(t_*) = -r, u''(t) < -r, \forall t \in (t_-, t_*)$.

Therefore, $\|u''\|_2 < \frac{R}{2} < R, \forall t \geq 0$.

Now, consider that $f$ verifies (1.2.3). By (1.2.5), consider that there are $t_-, t_* \in \mathbb{R}^+$ such that $t_- < t_*$ and $u''(t_*) = r, u''(t) > r, \forall t \in (t_-, t_*)$. Therefore, following similar steps as before,

$$\int_{u''(t_*)}^{u''(t_-)} \frac{s}{h(s)} ds = \int_{t_*}^{t_-} \frac{u''(s)}{h(u''(s))} u'''(s) ds \leq \int_{t_-}^{t_*} \psi(s) u''(s) ds$$

$$\leq \int_{t_-}^{t_*} \psi(s) u''(s) ds \leq M \int_{t_-}^{t_*} \frac{u''(s)}{(1+s)^\nu} ds$$

$$= M \left( M_1 + \sup_{t \geq 0} \frac{\Gamma'(t)}{1+t} \frac{\nu}{\nu-1} \right) < \int_r^R \frac{s}{h(s)} ds. \qquad (1.2.6)$$

So, $u''(t_-) < R$ and by the arbitrariness of $t_-$ and $t_*$ in $\mathbb{R}^+$, then $u''(t) < R, \forall t > 0$. The case where there are $t_*, t_+ \in \mathbb{R}^+$, with $t_* < t_+$, such that $u''(t_*) = -r, u''(t) < -r, \forall t \in (t_*, t_+]$ is proved in the same way. $\qquad \square$

The exact solution for the associated linear problem can be obtained by Green's functions method.

**Lemma 1.2.4.** *If $e \in L^1(\mathbb{R}_0^+)$, then the BVP*

$$\begin{cases} u'''(t) + e(t) = 0, & t \geq 0, \\ u(0) = A, \ au'(0) + bu''(0) = B, \ u''(+\infty) = C \end{cases} \qquad (1.2.7)$$

*has a unique solution in $X_1$. Moreover, this solution can be expressed as*

$$u(t) = g(t) + \int_0^{+\infty} G(t,s)e(s) ds, \qquad (1.2.8)$$

*where*

$$g(t) = \frac{Ct^2}{2} + \frac{B - bC}{a} t + A,$$

$$G(t,s) = \begin{cases} -\dfrac{b}{a} t + st - \dfrac{s^2}{2}, & 0 \leq s \leq t, \\ \dfrac{1}{2} t^2 - \dfrac{b}{a} t, & 0 \leq t \leq s < +\infty. \end{cases}$$

*Moreover,* $u'(t) = g'(t) + \int_0^{+\infty} G_1(t,s)e(s)ds$ *with*

$$G_1(t,s) = \begin{cases} -\dfrac{b}{a} + s, & 0 \le s \le t, \\[3mm] -\dfrac{b}{a} + t, & 0 \le t \le s < +\infty. \end{cases} \qquad (1.2.9)$$

The lack of compactness is overcome by the following lemma which gives a general criterium for relative compactness (see [4]).

**Lemma 1.2.5.** *A set $M \subset X_1$ is relatively compact if the following conditions hold:*

- (i) *all functions from $M$ are uniformly bounded;*
- (ii) *all functions from $M$ are equicontinuous on any compact interval of $\mathbb{R}_0^+$;*
- (iii) *all functions from $M$ are equiconvergent at infinity, that is, for any given $\epsilon > 0$, there exists a $t_\epsilon > 0$ such that*

$$\left| \frac{u^{(i)}(t)}{\omega_i(t)} - \frac{u^{(i)}(+\infty)}{\omega_i(+\infty)} \right| < \epsilon,$$

*for all $t > t_\epsilon$, $u \in M$ and $i = 0,1,2$.*

The well-known Schauder's fixed-point theorem will be the existence tool.

**Theorem 1.2.6 ([152]).** *Let $Y$ be a nonempty, closed, bounded and convex subset of a Banach space $X$, and suppose that $P : Y \to Y$ is a compact operator. Then $P$ is at least one fixed point in $Y$.*

An important tool to bound the solution and its derivatives is the lower and upper solution method. Let us define the usual lower and upper functions.

**Definition 1.2.7.** *Given $a > 0, b < 0$, and $A, B, C \in \mathbb{R}$, a function $\alpha \in C^3(\mathbb{R}_0^+) \cap X_1$ is a lower solution of problem (1.1.1),(1.1.2) if*

$$\begin{cases} \alpha'''(t) \ge f(t, \alpha(t), \alpha'(t), \alpha''(t)), & t \ge 0, \\[3mm] \alpha(0) \le A, \ a\alpha'(0) + b\alpha''(0) \le B, \ \alpha''(+\infty) < C. \end{cases}$$

A function $\beta \in C^3(\mathbb{R}_0^+) \cap X_1$ is an upper solution if it satisfies the reversed inequalities.

## 1.3. Existence and localization result

The main result of this chapter will be given by the following theorem.

**Theorem 1.3.1.** *Let* $f : \mathbb{R}_0^+ \times \mathbb{R}^3 \to \mathbb{R}$ *be an* $L^1$-*Carathéodory function. Suppose there are* $\alpha, \beta \in C^3(\mathbb{R}_0^+) \cap X_1$ *lower and upper solutions of the problem* (1.1.1),(1.1.2), *respectively, such that*

$$\alpha'(t) \leq \beta'(t), \quad \forall t \geq 0. \tag{1.3.1}$$

*If* $f$ *verifies either the one-sided Nagumo condition* (9.2.13) *or* (1.2.3) *in the set*

$$E_* = \left\{ (t, x, y, z) \in \mathbb{R}_0^+ \times \mathbb{R}^3, \alpha(t) \leq x \leq \beta(t), \alpha'(t) \leq y \leq \beta'(t) \right\},$$

*and*

$$f(t, \alpha(t), y, z) \geq f(t, x, y, z) \geq f(t, \beta(t), y, z), \tag{1.3.2}$$

*for* $(t, y, z)$ *fixed and* $\alpha(t) \leq x \leq \beta(t)$, *then the problem* (1.1.1),(1.1.2) *has at least one solution* $u \in C^3(\mathbb{R}_0^+) \cap X_1$ *and there exists* $R > 0$ *such that*

$$\alpha(t) \leq u(t) \leq \beta(t), \ \alpha'(t) \leq u'(t) \leq \beta'(t), \ \|u''\|_2 < R, \quad \forall t \geq 0.$$

**Remark 1.3.2.** By Theorem 1.3.1 and Definition 1.2.7, the following inequality is valid

$$\alpha(t) \leq \beta(t), \quad \forall t \geq 0,$$

and, therefore, $E_*$ is well defined and inequalities (1.3.2) make sense.

**Proof.** Let $\alpha, \beta \in C^3(\mathbb{R}_0^+) \cap X_1$ be, respectively, lower and upper solutions of (1.1.1),(1.1.2) verifying (1.3.1).

Consider the truncated and perturbed equation

$$u'''(t) = f\left(t, \delta_0(t), \delta_1(t), u''(t)\right) + \frac{1}{1+t^2} \frac{u'(t) - \delta_1(t)}{1 + |u'(t) - \delta_1(t)|}, \quad t \geq 0,$$

$$\tag{1.3.3}$$

where functions $\delta_j : \mathbb{R}_0^+ \times \mathbb{R} \to \mathbb{R}, j = 0, 1$ are given by

$$\delta_j(t) := \delta_j(t, u(t)) = \begin{cases} \beta^{(j)}(t), & u^{(j)}(t) > \beta^{(j)}(t), \\ u^{(j)}(t), & \alpha^{(j)}(t) \leq u^{(j)}(t) \leq \beta^{(j)}(t), \\ \alpha^{(j)}(t), & u^{(j)}(t) < \alpha^{(j)}(t). \end{cases} \qquad (1.3.4)$$

Note that the relation $\alpha(t) \leq \beta(t)$ is obtained by integration from (1.3.1) by the boundary conditions (1.1.2) and by Definition 1.2.7.

The proof will include three steps:

**Step 1:** *If $u$ is a solution of problem* (1.3.3),(1.1.2), *then*

$$\alpha(t) \leq u(t) \leq \beta(t), \ \alpha'(t) \leq u'(t) \leq \beta'(t), \quad \forall t \geq 0.$$

Suppose, by contradiction, that there exists $t \in \mathbb{R}_0^+$ with $\alpha'(t) > u'(t)$ and define

$$\inf_{t \geq 0}(u'(t) - \alpha'(t)) = u'(t_*) - \alpha'(t_*) < 0.$$

- If $t_* \in \mathbb{R}^+$, then $u''(t_*) = \alpha''(t_*)$ and $u'''(t_*) - \alpha'''(t_*) \geq 0$. Therefore, by (1.3.2) and Definition 1.2.7, the following contradiction holds:

$$0 \leq u'''(t_*) - \alpha'''(t_*)$$

$$= f(t_*, \delta_0(t_*), \delta_1(t_*), u''(t_*)) + \frac{1}{1 + t_*^2} \frac{u'(t_*) - \alpha'(t_*)}{1 + |u'(t_*) - \alpha'(t_*)|} - \alpha'''(t_*)$$

$$\leq f(t_*, \alpha(t_*), \alpha'(t_*), \alpha''(t_*)) + \frac{1}{1 + t_*^2} \frac{u'(t_*) - \alpha'(t_*)}{1 + |u'(t_*) - \alpha'(t_*)|} - \alpha'''(t_*)$$

$$\leq \frac{1}{1 + t_*^2} \frac{u'(t_*) - \alpha'(t_*)}{1 + |u'(t_*) - \alpha'(t_*)|} < 0.$$

- If $t_* = 0$, then

$$\min_{t \geq 0}(u'(t) - \alpha'(t)) := u'(0) - \alpha'(0) < 0,$$

and

$$u''(0) - \alpha''(0) \geq 0.$$

By Definition 1.2.7 and since $a > 0, b < 0$, it yields the contradiction

$$0 \geq bu''(0) - b\alpha''(0) \geq B - au'(0) - B + a\alpha'(0)$$
$$= a(\alpha'(0) - u'(0)) > 0.$$

- If $t_* = +\infty$, then

$$\inf_{t \geq 0}(u'(t) - \alpha'(t)) := u'(+\infty) - \alpha'(+\infty) < 0,$$

$$u''(+\infty) - \alpha''(+\infty) \leq 0,$$

and the following contradiction holds:

$$0 \geq u''(+\infty) - \alpha''(+\infty) > C - C = 0.$$

So, $\alpha'(t) \leq u'(t), \forall t \geq 0$. In a similar way, it can be proved that $\beta'(t) \geq u'(t), \forall t \geq 0$.

Integrating $\alpha'(t) \leq u'(t) \leq \beta'(t)$ on $[0, t]$ for $t \geq 0$, by (1.1.2) and Definition 1.2.7, it can be proved that $\alpha(t) \leq u(t) \leq \beta(t), \forall t \geq 0$.

**Step 2:** *If $u$ is a solution of the modified problem* (1.3.3),(1.1.2), *then there exists $R > 0$, not depending on $u$, such that*

$$\|u''\|_2 < R. \tag{1.3.5}$$

By the previous step, all solutions of equation (1.3.3) are solutions of (1.1.1), and as $f$ verifies either the one-sided Nagumo condition (9.2.13) or (1.2.3), this claim is a direct application of Lemma 1.2.3.

**Step 3:** *Problem* (1.3.3),(1.1.2) *has at least one solution.*
Take $\rho > \max\{\|\alpha\|_0, \|\beta\|_0, \|\alpha'\|_1, \|\beta'\|_1, R\}$ with $R$ given by (1.3.5).
Define the operator $T : X_1 \to X_1$ given by

$$Tu(t) = g(t) + \int_0^{+\infty} G(t, s)F(u(s))ds$$

with

$$g(t) := \frac{C}{2}t^2 + \frac{B - bC}{a}t + A$$

and

$$F(u(s)) := f(s, \delta_0(s), \delta_1(s), u''(s)) + \frac{1}{1 + s^2}\frac{u'(s) - \delta_1(s)}{1 + |u'(s) - \delta_1(s)|}.$$

As $f$ is a $L^1$-Carathéodory function, for any $u \in X_1$ with $\|u\|_{X_1} < \rho$, then $F \in L^1$ because

$$\int_0^{+\infty} |F(u(s))| \, ds \leq \int_0^{+\infty} \varphi_\rho(s) + \frac{1}{1 + s^2}\frac{|u'(s) - \delta_1(s)|}{1 + |u'(s) - \delta_1(s)|} \, ds$$

$$\leq \int_0^{+\infty} \varphi_\rho(s) + \frac{1}{1 + s^2} \, ds < +\infty. \tag{1.3.6}$$

By Lemma 8.2.1, the fixed points of $T$ are solutions of problem (1.3.3),(1.1.2). So it is enough to prove that $T$ has a fixed point.

**Claim 1.** $T : X_1 \to X_1$ *is well defined.*
By the Lebesgue Dominated Theorem and Lemma 1.2.4,

$$\lim_{t \to +\infty} \frac{(Tu)(t)}{1 + t^2} \leq \frac{C}{2} + \frac{1}{2}\int_0^{+\infty} F(u(s)) \, ds < +\infty.$$

Analogously, by (1.2.9),

$$\lim_{t \to +\infty} \frac{(Tu)'(t)}{1 + t} = \lim_{t \to +\infty} \frac{g'(t)}{1 + t} + \int_0^{+\infty} \lim_{t \to +\infty} \frac{G_1(t, s)}{1 + t} F(u(s)) \, ds$$

$$\leq C + \int_0^{+\infty} F(u(s)) \, ds < +\infty,$$

and

$$\lim_{t \to +\infty} \frac{(Tu)''(t)}{2} \leq \frac{C}{2} + \frac{1}{2}\lim_{t \to +\infty}\int_t^{+\infty} F(u(s)) \, ds = \frac{C}{2} < +\infty.$$

Therefore, $Tu \in X_1$.

**Claim 2.** *T is continuous.*

Consider a convergent sequence $u_n \to u$ in $X_1$. Then there exists $r_1 > 0$ such that $\|u_n\|_{X_1} < r_1$ and

$$\|Tu_n - Tu\|_{X_1} \leq \int_0^{+\infty} \max \left\{ \begin{array}{c} \sup_{t \geq 0} \left| \dfrac{G(t,s)}{1+t^2} \right|, \\[3mm] \sup_{t \geq 0} \left| \dfrac{G_1(t,s)}{1+t} \right|, \dfrac{1}{2} \end{array} \right\}$$

$$\times |F(u_n(s)) - F(u(s))| ds$$

$$\leq \int_0^{+\infty} |F(u_n(s)) - F(u(s))| \, ds \longrightarrow 0, \qquad (1.3.7)$$

as $n \to +\infty$.

**Claim 3.** *T is compact.*

Let

$$M(s) := \max \left\{ \sup_{t \geq 0} \frac{|G(t,s)|}{1+t^2}, \sup_{t \geq 0} \frac{|G_1(t,s)|}{1+t} \right\}.$$

Consider a bounded set $B \subset X_1$ defined by $B := \{u \in X_1 : \|u\|_{X_1} < r_1\}$ for some $r_1 > 0$ such that

$$r_1 > \max \left\{ \rho, \frac{|C|}{2} + \int_0^{+\infty} M(s) \left( \varphi_\rho(s) + \frac{1}{1+s^2} \right) ds \right\}$$

with $\rho$ given by (1.3.6).

**Claim 3.1.** *TB is uniformly bounded.*

For any $u \in B$, as $\|\alpha\|_0 \leq \|\delta_0\|_0 \leq \|\beta\|_0$, $\|\alpha'\|_1 \leq \|\delta_1\|_1 \leq \|\beta'\|_1$, by (1.2.2), one has

$$\|Tu\|_0 = \sup_{t \geq 0} \frac{|Tu(t)|}{1+t^2} \leq \sup_{t \geq 0} \frac{|g(t)|}{1+t^2} + \int_0^{+\infty} \sup_{t \geq 0} \frac{|G(t,s)|}{1+t^2} |F(u(s))| \, ds$$

$$\leq \frac{|C|}{2} + \int_0^{+\infty} M(s) \left( \varphi_\rho(s) + \frac{1}{1+s^2} \right) ds < r_1,$$

$$\|Tu\|_1 = \sup_{t \geq 0} \frac{|(Tu)'(t)|}{1+t} \leq \sup_{t \geq 0} \frac{|g'(t)|}{1+t} + \int_0^{+\infty} \sup_{t \geq 0} \frac{|G_1(t,s)|}{1+t} |F(u(s))| \, ds$$

$$\leq |C| + \int_0^{+\infty} M(s) \left( \varphi_\rho(s) + \frac{1}{1+s^2} \right) ds < r_1,$$

and

$$\|Tu\|_2 = \sup_{t \geq 0} \frac{|(Tu)''(t)|}{2} \leq \frac{|C|}{2} < r_1.$$

Thus, $\|Tu\|_{X_1} < r_1$, $TB$ is uniformly bounded, and, moreover, $TB \subset B$.

**Claim 3.2.** *TB is equicontinuous.*
For $T > 0$ and $t_1, t_2 \in [0, T]$,

$$\left| \frac{Tu(t_1)}{1 + t_1^2} - \frac{Tu(t_2)}{1 + t_2^2} \right| \leq \left| \frac{g(t_1)}{1 + t_1^2} - \frac{g(t_2)}{1 + t_2^2} \right|$$
$$+ \int_0^{+\infty} \left| \frac{G(t_1, s)}{1 + t_1^2} - \frac{G(t_2, s)}{1 + t_2^2} \right|$$
$$\times |F(u(s))| \, ds \longrightarrow 0, \ as \ t_1 \to t_2.$$

Analogously,

$$\left| \frac{(Tu)'(t_1)}{1 + t_1} - \frac{(Tu)'(t_2)}{1 + t_2} \right| = \left| \frac{g'(t_1)}{1 + t_1} - \frac{g'(t_2)}{1 + t_2} \right|$$
$$+ \int_0^{+\infty} \left| \frac{G_1(t_1, s)}{1 + t_1} - \frac{G_1(t_2, s)}{1 + t_2} \right|$$
$$\times |F(u(s))| \, ds \longrightarrow 0, \ as \ t_1 \to t_2,$$

and

$$\left| \frac{(Tu)''(t_1)}{2} - \frac{(Tu)''(t_2)}{2} \right| = \left| \int_{t_1}^{t_2} F(s) ds \right|$$
$$\leq \int_{t_1}^{t_2} \left( \varphi_\rho(s) + \frac{1}{1 + s^2} \right) ds \longrightarrow 0, \ as \ t_1 \to t_2.$$

**Claim 3.3.** *TB is equiconvergent at infinity.* Indeed,

$$\left| \frac{Tu(t)}{1 + t^2} - \lim_{t \to +\infty} \frac{Tu(t)}{1 + t^2} \right| \leq \left| \frac{g(t)}{1 + t^2} - \frac{C}{2} \right|$$
$$+ \int_0^{+\infty} \left| \frac{G(t, s)}{1 + t^2} - \lim_{t \to +\infty} \frac{G(t, s)}{1 + t^2} \right|$$
$$\times |F(u(s))| \, ds \longrightarrow 0, \ as \ t \to +\infty,$$

$$\left| \frac{(Tu)'(t)}{1+t} - \lim_{t \to +\infty} \frac{(Tu)'(t)}{1+t} \right| \le \left| \frac{g'(t)}{1+t} - C \right|$$

$$+ \int_0^{+\infty} \left| \frac{G_1(t,s)}{1+t} - \lim_{t \to +\infty} \frac{G_1(t,s)}{1+t} \right|$$

$$\times |F(u(s))| ds \longrightarrow 0, as \ t \to +\infty,$$

and

$$\left| \frac{(Tu)''(t)}{2} - \lim_{t \to +\infty} (Tu)''(t) \right| = \left| \int_t^{+\infty} F(u(s)) ds \right|$$

$$\le \int_t^{+\infty} \left( \varphi_\rho(s) + \frac{1}{1+s^2} \right)$$

$$\times ds \longrightarrow 0, \quad as \ t \to +\infty.$$

So, by Lemma 1.2.5, $TB$ is relatively compact.

As $T$ is completely continuous, then by Schauder's fixed-point theorem (Theorem 1.2.6), $T$ has at least one fixed point $u \in X_1$. □


## 1.4. Example

Consider the next third-order BVP

$$\begin{cases} u'''(t) = \dfrac{1}{(t+1)^2}(-\arctan(u(t)) - 10|u''(t)|e^{u''(t)}), & t \ge 0, \\[2mm] u(0) = A, \ au'(0) + bu''(0) = B, \ u''(+\infty) = C, \end{cases} \tag{1.4.1}$$

with $A \in (-1,0]$, $a > 0$, $b < 0$ such that $-2(a+b) \le B \le 0$ and $C \in (-2,0)$.
Define

$$E_{ex1} = \left\{ (t,x,y,z) \in \mathbb{R}_0^+ \times \mathbb{R}^3 : -(t+1)^2 \le x \le 0, -2t-2 \le y \le 0 \right\}.$$

Function $f : \mathbb{R}_0^+ \times \mathbb{R}^3 \to \mathbb{R}$ defined by

$$f(t,x,y,z) := \frac{1}{(t+1)^2} \left( -\arctan x - 10|z|e^z \right)$$

verifies on $E_{ex1}$ the inequality $|f(t,x,y,z)| \le \frac{K_\rho}{(t+1)^2} := \varphi_\rho(t)$ for some $K_\rho > 0$ and $\rho$ such that $\max\{2, \|z\|_2\} < \rho$. Therefore, $f$ is $L^1$-Carathéodory.

Functions $\alpha(t) = -(t+1)^2$ and $\beta(t) \equiv 0$ are, respectively, lower and upper solutions of problem (1.4.1) with $\alpha(t) \le \beta(t)$ and $\alpha'(t) \le \beta'(t)$, $\forall t \ge 0$, verifying (1.3.2).

As

$$f(t,x,y,z) \leq \frac{1}{(t+1)^2}\frac{\pi}{2},$$

the one-sided Nagumo-type growth condition (9.2.13) holds in $E_{ex1}$ with

$$\psi(t) := \frac{1}{(t+1)^2}, \ \nu \in (1,2), \text{ and } h(|z|) := \frac{\pi}{2}.$$

Therefore, by Theorem 1.3.1, there is at least a solution $u$ of (1.4.1) with

$$-(t+1)^2 \leq u(t) \leq 0, \ -2t - 2 \leq u'(t) \leq 0, \ \|u''\|_2 < R, \ \forall t \geq 0.$$

Moreover, from the localization part of the theorem, one can express some qualitative properties of this solution: it is nonpositive, nonincreasing and, as $C \neq 0$, this solution is unbounded.

Note that $f$ does not satisfy the usual two-sided Nagumo-type condition. In fact, if there exist $\psi_1, h_1 \in C(\mathbb{R}_0^+, \mathbb{R}^+)$ satisfying

$$|f(t,x,y,z)| \leq \psi_1(t)\, h_1(|z|), \quad \forall (t,x,y,z) \in E_{ex1},$$

with $\int_0^{+\infty} \frac{s}{h_1(s)} ds = +\infty$, then, in particular,

$$-f(t,x,y,z) \leq \psi_1(t)\, h_1(|z|), \quad \forall (t,x,y,z) \in E_{ex1}.$$

So, for $x = 0, \ y, z \in \mathbb{R}$, one has

$$-f(t,0,y,z) = \frac{10}{(t+1)^2}|z|e^z \leq \psi_1(t)\, h_1(|z|).$$

Considering $\psi_1(t) := \frac{1}{(t+1)^2}$ , the following contradiction holds:

$$+\infty > \int_0^{+\infty} \frac{s}{10se^s} ds \geq \int_0^{+\infty} \frac{s}{h_1(s)} ds = +\infty.$$

## Chapter 2

# General $n$th-Order Problems

## 2.1. Introduction

As shown in Chapter 1, $n$th-Order BVPs on infinite intervals occur in different areas. For example, fourth-order differential equations can model the bending of an elastic beam and, in this sense, they are called beam equations. Other higher order problems are related with the study of radially symmetric solutions of nonlinear elliptic equations, fluid dynamics, boundary layer theory, semiconductor circuits and soil mechanics, either on the bounded domains (see [12, 40, 62, 120]) or on the real line ( [3, 50, 86, 87, 100]).

The study of BVPs on bounded domains is vast, but focus on infinite intervals is scarce. Different methods, such as fixed point theorems, shooting methods, upper and lower technique, are used to prove the existence of solutions. However, these solutions are usually bounded.

Lower and upper solutions method, coupled with the Nagumo-type condition, guarantees the existence of at least one solution lying on the strip defined by lower and upper solutions (see [100]) but, to the best of our knowledge, there are no results when the nonlinearity satisfies only the one-sided Nagumo-type condition on unbounded intervals.

This chapter concerns the study of a general Sturm–Liouville-type BVP composed by the $n$th-order differential equation defined on the half-line

$$u^{(n)}(t) = f(t, u(t), u'(t), \ldots, u^{(n-1)}(t)), \quad \text{a.e. } t \geq 0, \qquad (2.1.1)$$

and

$$\begin{cases} u^{(i)}(0) = A_i, \\ u^{(n-2)}(0) + au^{(n-1)}(0) = B, \\ u^{(n-1)}(+\infty) = C, \end{cases} \tag{2.1.2}$$

with $f : \mathbb{R}_0^+ \times \mathbb{R}^n \to \mathbb{R}$ an $L^1$-Carathéodory function, $a < 0, A_i, B, C \in \mathbb{R}$ for $i = 0, 1, \ldots, n-3$, and $u^{(n-1)}(+\infty) := \lim_{t \to +\infty} u^{(n-1)}(t)$.

The functional setting will be adapted to the $n$th-order case, namely, the weight space, the corresponding norms and the notion of $L^1$-Carathéodory.

As an application of this result, we include a particular case of a fourth-order problem with a beam equation, referred to in [46].

## 2.2. Preliminary results

A new admissible space will be needed.

For polynomial functions $\omega_i(t) = 1 + t^{n-1-i}, i = 0, 1, \ldots, n-1$, let us define the space

$$X_2 = \left\{ x \in C^{n-1}(\mathbb{R}_0^+) : \lim_{t \to +\infty} \frac{x^{(i)}(t)}{\omega_i(t)} \in \mathbb{R}, i = 0, 1, \ldots, n-1 \right\},$$

with the norm $\|x\|_{X_2} = \max\left\{ \|x\|_0, \|x'\|_1, \ldots, \|x^{(n-1)}\|_{n-1} \right\}$, where

$$\|y\|_i = \sup_{t \geq 0} \left| \frac{y(t)}{\omega_i(t)} \right|, \quad \text{for } i = 0, 1, \ldots, n-1.$$

It is clear that $(X_2, \| \cdot \|_{X_2})$ is a Banach space.

Let $\gamma_i, \Gamma_i \in C(\mathbb{R}_0^+), \gamma_i(t) \leq \Gamma_i(t), \forall t \geq 0, \; i = 0, 1, \ldots, n-2$ and define

$$E_2 = \{(t, x_0, \ldots, x_{n-1}) \in \mathbb{R}_0^+ \times \mathbb{R}^n : \gamma_i(t) \leq x_i \leq \Gamma_i(t), i = 0, 1, \ldots, n-2\}.$$

Now, the one-sided growth condition can be formulated in the following way.

**Definition 2.2.1.** A function $f : E_2 \to \mathbb{R}$ is said to satisfy a one-sided Nagumo-type growth condition in $E_2$ if, for some positive and continuous functions $\psi, h$ and some $\nu > 1$, such that

$$\int_0^{+\infty} \psi(s)ds < +\infty, \; \sup_{t \geq 0} \psi(t)(1+t)^\nu < +\infty, \; \int_0^{+\infty} \frac{s}{h(s)}ds = +\infty,$$

(2.2.1)

it verifies either

$$f(t, x_0, \ldots, x_{n-1}) \leq \psi(t)h(\|x_{n-1}\|_{n-1}), \quad \forall (t, x_0, \ldots, x_{n-1}) \in E_2$$

(2.2.2)

or

$$f(t, x_0, \ldots, x_{n-1}) \geq -\psi(t)h(\|x_{n-1}\|_{n-1}), \quad \forall (t, x_0, \ldots, x_{n-1}) \in E_2.$$

(2.2.3)

Now, the *a priori* estimation is obtained on $u^{(n-1)}$, given by the following lemma, where the proof follows the same technique as in Lemma 1.2.3 and, for this reason, is omitted.

**Lemma 2.2.2.** *Let $f : \mathbb{R}_0^+ \times \mathbb{R}^n \to \mathbb{R}$ be an $L^1$-Carathéodory function satisfying (2.2.1) and (2.2.2), or (2.2.3), in $E_2$. Then there exists $R > 0$ (not depending on $u$) such that every $u$ solution of (2.1.1), (2.1.2) satisfying*

$$\gamma_i(t) \leq u^{(i)}(t) \leq \Gamma_i(t), \quad \forall t \geq 0, \; i = 0, 1, \ldots, n-2 \quad (2.2.4)$$

*verifies $\left\|u^{(n-1)}\right\|_{n-1} < R$.*

The exact solution for the associated linear problem can be obtained by a Green function.

**Lemma 2.2.3.** *If $e \in L^1(\mathbb{R}_0^+)$, then the BVP*

$$\begin{cases} u^{(n)}(t) + e(t) = 0, \; a.e. \; t \geq 0, \\ u^{(i)}(0) = A_i, \quad i = 0, 1, \ldots, n-3, \\ u^{(n-2)}(0) + au^{(n-1)}(0) = B, \\ u^{(n-1)}(+\infty) = C \end{cases}$$

(2.2.5)

*has a unique solution in $X_2$. Moreover, this solution can be expressed as*

$$u(t) = p(t) + \int_0^{+\infty} G(t, s)e(s)ds,$$

(2.2.6)

*where*

$$p(t) = \sum_{k=0}^{n-3} \frac{A_k}{k!} t^k + \frac{B - aC}{(n-2)!} t^{n-2} + \frac{C}{(n-1)!} t^{n-1},$$

*and*

$G(t, s)$

$$= \begin{cases} \displaystyle\sum_{k=0}^{n-2} \frac{(-1)^k}{(k+1)!(n-2-k)!} s^{k+1} t^{n-2-k} - \frac{at^{n-2}}{(n-2)!}, & 0 \leq s \leq t < +\infty, \\[3ex] \dfrac{1}{(n-1)!} t^{n-1} - \dfrac{a}{(n-2)!} t^{n-2}, & 0 \leq t \leq s < +\infty. \end{cases}$$

General $n$th-order definitions of lower and upper functions are presented next.

**Definition 2.2.4.** Given $a < 0$ and $A_i, B, C \in \mathbb{R}, i = 0, 1, \ldots, n-3$, a function $\alpha \in C^n(\mathbb{R}_0^+) \cap X_2$ is a lower solution of problem (2.1.1),(2.1.2) if

$$\begin{cases} \alpha^{(n)}(t) \geq f(t, \alpha(t), \alpha'(t), \ldots, \alpha^{(n-1)}(t)), & t \geq 0, \\ \alpha^{(i)}(0) \leq A_i, \\ \alpha^{(n-2)}(0) + a\alpha^{(n-1)}(0) \leq B, \\ \alpha^{(n-1)}(+\infty) < C. \end{cases}$$

A function $\beta \in C^n(\mathbb{R}_0^+) \cap X_2$ is an upper solution if it satisfies the reversed inequalities.

## 2.3.   Existence and localization result

The existence theorem to the $n$th-order case follows similar arguments of Theorem 1.3.1, and the proof is omitted.

**Theorem 2.3.1.** *Let $f : \mathbb{R}_0^+ \times \mathbb{R}^n \to \mathbb{R}$ be an $L^1$-Carathéodory function. Suppose there are $\alpha, \beta \in C^n(\mathbb{R}_0^+) \cap X_2$, lower and upper solutions of the problem (2.1.1),(2.1.2), respectively, such that*

$$\alpha^{(n-2)}(t) \leq \beta^{(n-2)}(t), \quad \forall t \geq 0. \tag{2.3.1}$$

*If f verifies either the one-sided Nagumo condition (2.2.2) or (2.2.3) in the set*

$$E_* = \{(t, x_0, \ldots, x_{n-1}) \in \mathbb{R}_0^+ \times \mathbb{R}^n : \alpha^{(i)}(t) \le x_i \le \beta^{(i)}(t), i = 0, \ldots, n-2\},$$

*and*

$$\begin{aligned}
&f(t, \alpha(t), \ldots, \alpha^{(i)}(t), \ldots, u_{n-2}, u_{n-1}) \\
&\ge f(t, u_0, \ldots, u_i, \ldots, u_{n-2}, u_{n-1}) \\
&\ge f(t, \beta(t), \ldots, \beta^{(i)}(t), \ldots, u_{n-2}, u_{n-1}),
\end{aligned} \qquad (2.3.2)$$

*for $(t, u_{n-2}, u_{n-1})$ fixed when $\alpha^{(i)}(t) \le u_i \le \beta^{(i)}(t), i = 0, \ldots, n-3$, then problem (2.1.1),(2.1.2) has at least one solution $u \in C^n(\mathbb{R}_0^+) \cap X_2$ and there exists $R > 0$ such that*

$$\alpha^{(i)}(t) \le u^{(i)}(t) \le \beta^{(i)}(t), \quad i = 0, 1, \ldots, n-2 \quad \text{and}$$

$$\|u^{(n-1)}\|_{n-1} < R, \ \forall t \ge 0.$$

**Remark 2.3.2.** Note that by integration on $[0, t]$ of (2.3.1) and Definition 2.2.4, lower and upper solutions and their derivatives (until order $n-3$) are well ordered, that is,

$$\alpha^{(i)}(t) \le \beta^{(i)}(t), \quad i = 0, 1, \ldots, n-3, \ \forall t \ge 0,$$

and $E_*$ is well defined.

## 2.4.  Example

Consider the next fourth-order BVP

$$\begin{cases}
u^{(iv)}(t) = \dfrac{-u(t)|u'''(t) - 6|e^{u'''(t)} - e^{-t}(6t + 2 - u''(t))}{1 + t^2}, & t \ge 0, \\
u(0) = A, \ u'(0) = 0, \ u''(0) + au'''(0) = 0, \ u'''(+\infty) = C,
\end{cases}$$

$$(2.4.1)$$

with $A \ge 0$, $-\frac{1}{3} \le a < 0$ and $0 < C < 6$.

This BVP is a particular case of (2.1.1), (2.1.2) with $A_0 = A, A_1 = 0$, $B = 0$ and

$$f(t, x, y, z, w) = \frac{-x|w - 6|e^w - e^{-t}(6t + 2 - z)}{1 + t^2}. \qquad (2.4.2)$$

Moreover, functions $\alpha(t) \equiv A$ and $\beta(t) = t^3 + t^2 + A$ are, respectively, lower and upper solutions for (10.3.1), and Nagumo condition with (2.2.2) is verified with

$$\psi(t) = \frac{1}{1+t^2}, \quad 1 < \nu < 2, h(|w|) \equiv 1,$$

on

$$E_{ex2} = \left\{ (t,x,y,z,w) \in \mathbb{R}_0^+ \times \mathbb{R}^4 : \begin{array}{c} A \le x \le t^3 + t^2 + A \\ 0 \le y \le 3t^2 + 2t \\ 0 \le z \le 6t + 2 \end{array} \right\}.$$

Also, $f$ verifies (2.3.2) and all assumptions of Theorem 2.3.1 are fulfilled, therefore, there is at least a nontrivial solution $u$ of (2.4.1) such that

$$A \le u(t) \le t^3 + t^2 + A,$$
$$0 \le u'(t) \le 3t^2 + 2t,$$
$$0 \le u''(t) \le 6t + 2,$$
$$\|u'''\|_3 \le R, \quad \forall t \ge 0.$$

Remark that this solution is unbounded and, from the location part, it is nondecreasing and convex.

It is important to stress that the nonlinearity (2.4.2) does not satisfy the usual two-sided Nagumo-type condition. Therefore, the existent results in the literature cannot be applied to problem (2.4.1).

In fact, if there exist $\psi_2, h_2 \in C(\mathbb{R}_0^+, \mathbb{R}^+)$ satisfying

$$|f(t,x,y,z,w)| \le \psi_2(t)h_2(|w|), \quad \forall (t,x,y,z,w) \in E_{ex2},$$

with $\int_0^{+\infty} \frac{s}{h_2(s)} ds = +\infty$, then, in particular,

$$-f(t,x,y,z,w) \le \psi_2(t)h_2(|w|),$$

and, for $t \ge 0$, $x = 1$, $0 \le y \le 3t^2 + 2t$, $z = 6t + 2$, and $w \in \mathbb{R}$,

$$-f(t,1,y,6t+2,w) = \frac{|w - 6|e^w}{1+t^2} \le \psi_2(t)h_2(|w|).$$

For $\psi_2(t) = \frac{1}{1+t^2}$, one has $|w - 6|e^w \le h_2(|w|)$ and the following contradiction holds:

$$+\infty > \int_0^{+\infty} \frac{s}{(s-6)e^s} ds \ge \int_0^{+\infty} \frac{s}{h_2(s)} ds = +\infty.$$

# Chapter 3

# Impulsive Problems on the Half-Line with Infinite Impulse Moments

## 3.1. Introduction

This chapter concerns the following boundary value problem composed by the differential equation

$$u''(t) = f(t, u(t), u'(t)), \quad \text{a.e. } t \in ]0, +\infty[, \ t \neq t_k, \quad (3.1.1)$$

where $f : [0, +\infty[ \times \mathbb{R}^2 \to \mathbb{R}$ is an $L^1$-Carathéodory function, the two-point boundary conditions on the half-line

$$u(0) = A,$$
$$u'(+\infty) = B, \quad (3.1.2)$$

with $A, B \in \mathbb{R}$, $u'(+\infty) := \lim_{t \to +\infty} u'(t)$, and the impulsive effects

$$\Delta u(t_k) = I_{0k}(t_k, u(t_k), u'(t_k)),$$
$$\Delta u'(t_k) = I_{1k}(t_k, u(t_k), u'(t_k)), \quad (3.1.3)$$

where $k \in \mathbb{N}$, $\Delta u^{(i)}(t_k) = u^{(i)}(t_k^+) - u^{(i)}(t_k^-)$, $I_{ik} \in C([0, +\infty[ \times \mathbb{R}^2, \mathbb{R})$, $i = 0, 1$, and $I_{0k}$ with $t_k$ fixed points such that $0 = t_0 < t_1 < t_2 < \cdots < t_k < \cdots$ and $\lim_{k \to +\infty} t_k = +\infty$.

Impulsive boundary value problems (IBVP) of different types have been the object of increasing attention (see, for example, [21, 34, 56, 66, 67, 104, 116, 129, 131]) as they are well adapted to describe real phenomena where a sudden change of their state occurs at certain moments. These situations often happen in physics, chemistry, population dynamics, biotechnology, economics and control theory, among others (see [20, 95] and the references therein).

In the recent years, these problems had also been considered on unbounded domains with finite or infinite impulsive instants, applying different methods to deal with the lack of compactness: variational techniques, lower and upper solutions, coincidence degree theory and fixed point theorems on adequate Banach spaces (see, for instance, [54, 101, 145]).

Motivated by these works, we consider problem (3.1.1)–(3.1.3). To the author's best knowledge, it is the first time where the second-order IBVP is considered in the half-line with general nonlinearity and with infinite impulsive effects, on the unknown function and its first derivative, given by generalized functions. Therefore, this problem can model cases where the occurrence of infinite jumps depends not only on the instant, but also on their amplitude and frequency.

The arguments are applied in an adequate Banach space defined with weighted norms with Green's functions to obtain an integral operator and Schauder's fixed-point theorem. We point out that the equiconvergence at $+\infty$ and at each impulsive moment is a key point to have a compact operator. Moreover, the notion of Carathéodory sequences is useful to control the behavior of the impulsive functions. In this way, no other assumptions, such as sublinearity, superlinearity or monotone types, are needed.

## 3.2.   Definitions and preliminary results

This section contains some definitions and auxiliary results used along the chapter.

For $u(t_k^{\pm}) := \lim_{t \to t_k^{\pm}} u(t)$, consider the sets

$$PC\left([0,+\infty[\right) = \left\{ \begin{array}{l} u : u \in C([0,+\infty[, \mathbb{R}) \\ \text{continuous for } t \neq t_k, u(t_k) = u(t_k^-) \\ u(t_k^+) \text{ exists for } k \in \mathbb{N} \end{array} \right\},$$

$PC^1\left([0,+\infty[\right) = \{u : u'(t) \in PC\left([0,+\infty[\right)\}$, and the space

$$X = \left\{ x \in PC^1\left([0,+\infty[\right) : \lim_{t \to +\infty} \frac{x(t)}{1+t} \in \mathbb{R}, \lim_{t \to +\infty} x'(t) \in \mathbb{R} \right\}.$$

Defining the norm $\|x\|_X = \max\{\|x\|_0, \|x'\|_1\}$, where

$$\|\omega\|_0 := \sup_{0 \le t < +\infty} \frac{|\omega(t)|}{1+t} \quad \text{and} \quad \|\omega\|_1 := \sup_{0 \le t < +\infty} |\omega(t)|,$$

then $(X, \|\cdot\|_X)$ is a Banach space.

The function is a solution $u$ of problem (3.1.1)–(3.1.3) if $u(t) \in X$ and verifies conditions (3.1.1)–(3.1.3).

**Definition 3.2.1.** A sequence $(w_n)_{n \in \mathbb{N}} : [0, +\infty[ \times \mathbb{R}^2 \to \mathbb{R}$ is a Carathéodory sequence if it verifies the following conditions:

(i) for each $u, v \in \mathbb{R}$, $(u, v) \to w_n(t, u, v)$ is continuous for all $n \in \mathbb{N}$;

(ii) for each $\rho > 0$, there are nonnegative constants $\Psi_{n,\rho} \ge 0$ with $\sum_{n=1}^{+\infty} \Psi_{n,\rho} < +\infty$ such that for $|u| < \rho(1+t)$, $t \in [0, +\infty[$, $|v| < \rho$ we have

$$|w_n(t, u, v)| \le \psi_{n,\rho}, \quad \text{for every } n \in \mathbb{N}, t \in [0, +\infty[.$$

For a linear problem associated with the initial one, we have the following uniqueness result obtained via Green's functions by standard techniques.

**Lemma 3.2.2.** *Let* $h : [0, +\infty[ \to \mathbb{R}$ *be an* $L^1$-*Carathéodory function and* $I_{1k} : [0, +\infty[ \times \mathbb{R}^2 \to \mathbb{R}$ *be a Carathéodory sequence. Then the problem composed by the differential equation*

$$u''(t) = h(t), \quad \text{a.e.} \quad t \in [0, +\infty[, \tag{3.2.1}$$

*and conditions* (3.1.2), (3.1.3), *has a unique solution defined by*

$$u(t) = A + Bt + \sum_{0 < t_k < t < +\infty} [I_{0k}(t_k, u(t_k), u'(t_k))$$

$$+ I_{1k}(t_k, u(t_k), u'(t_k))(t - t_k)]$$

$$- t \sum_{k=1}^{+\infty} I_{1k}(t_k, u(t_k), u'(t_k)) + \int_0^{+\infty} G(t, s) \, h(s) \, ds, \tag{3.2.2}$$

*where*

$$G(t,s) = \begin{cases} -s, & 0 \le s \le t, \\ -t, & t \le s < +\infty. \end{cases}$$

The next lemma provides a general criterion for relative compactness on $X$.

**Lemma 3.2.3 ([3]).** *A set $M \subset X$ is relatively compact if the following conditions hold:*

(i) *all functions from $M$ are uniformly bounded;*
(ii) *all functions from $M$ are equicontinuous on any compact interval of $[0, +\infty[$;*
(iii) *all functions from $M$ are equiconvergent at infinity, that is, for any given $\epsilon > 0$, there exists a $t_\epsilon > 0$ such that*

$$\left| \frac{x(t)}{1+t} - \lim_{t \to +\infty} \frac{x(t)}{1+t} \right| < \epsilon, \left| x'(t) - \lim_{t \to +\infty} x'(t) \right| < \epsilon \text{ for all } t > t_\epsilon, x \in M.$$

## 3.3.   Main result

In this section, sufficient conditions are given for the solvability of problems (3.1.1)–(3.1.3).

**Theorem 3.3.1.** *Let $f : [0, +\infty[ \times \mathbb{R}^2 \to \mathbb{R}$ be an $L^1$-Carathéodory function. If $I_{0k}, I_{1k} : [0, +\infty[ \times \mathbb{R}^2 \to \mathbb{R}$ are Carathéodory sequences with nonnegative constants $\varphi_{k,\rho} \ge 0$, $\psi_{k,\rho} \ge 0$ with $\sum_{k=1}^{+\infty} \varphi_{k,\rho} < +\infty$, $\sum_{k=1}^{+\infty} \psi_{k,\rho} < +\infty$, such that*

$$|I_{0k}(t_k, x, y)| \le \varphi_{k,\rho}, \quad |I_{1k}(t_k, x, y)| \le \psi_{k,\rho}, \tag{3.3.1}$$

*for $|x| < \rho(1+t)$, $t \in [0, +\infty[$, $|y| < \rho$, then problem (3.1.1)–(3.1.3) has at least a solution $u \in X$.*

**Proof.** Define the operator $T : X \to X$

$$Tu(t) = A + Bt + \sum_{0 < t_k < t < +\infty} [I_{0k}(t_k, u(t_k), u'(t_k))$$

$$+ I_{1k}(t_k, u(t_k), u'(t_k))(t - t_k)]$$

$$- t \sum_{k=1}^{+\infty} I_{1k}(t_k, u(t_k), u'(t_k)) + \int_0^{+\infty} G(t,s) f(s, u(s), u'(s)) ds.$$

By Lemma 3.2.2, a fixed point of $T$ is a solution of problem (3.1.1)–(3.1.3). The proof that operator $T$ has a fixed point will follow several steps.

**Step 1.** $T$ *is well defined and continuous on* $X$.

As $f$ is an $L^1$-Carathéodory function, $Tu \in PC^1([0, +\infty[)$. Moreover, by the Lebesgue Dominated Convergence Theorem,

$$
\begin{aligned}
\lim_{t \to +\infty} \frac{(Tu)(t)}{1+t} &= \lim_{t \to +\infty} \frac{A + Bt}{1+t} \\
&\quad + \frac{1}{1+t} \sum_{0 < t_k < t < +\infty} [I_{0k}(t_k, u(t_k), u'(t_k)) \\
&\quad + I_{1k}(t_k, u(t_k), u'(t_k))(t - t_k)] \\
&\quad - \frac{t}{1+t} \sum_{k=1}^{+\infty} I_{1k}(t_k, u(t_k), u'(t_k)) \\
&\quad + \int_0^{+\infty} \lim_{t \to +\infty} \frac{G(t,s)}{1+t} f(s, u(s), u'(s)) ds \\
&= B + \sum_{0 < t_k < t < +\infty} I_{1k}(t_k, u(t_k), u'(t_k)) \\
&\quad - \sum_{k=1}^{+\infty} I_{1k}(t_k, u(t_k), u'(t_k)) - \int_0^{+\infty} f(s, u(s), u'(s)) ds \\
&\leq B + 2 \sum_{k=1}^{+\infty} \psi_{k,\rho_1} + \int_0^{+\infty} \varphi_\rho(s) ds < +\infty,
\end{aligned}
$$

and

$$
\begin{aligned}
\lim_{t \to +\infty} (Tu)'(t) &= B + \sum_{k=1}^{+\infty} I_{1k}(t_k, u(t_k), u'(t_k)) - \sum_{k=1}^{+\infty} I_{1k}(t_k, u(t_k), u'(t_k)) \\
&\quad - \lim_{t \to +\infty} \int_t^{+\infty} f(s, u(s), u'(s)) ds \\
&= B < +\infty.
\end{aligned}
$$

Therefore, $Tu \in X$.

**Step 2.** *TD is uniformly bounded for D any bounded set on X.*

Let $D \subset X$ be a bounded subset on $X$. So, there exists $\rho_1 > 0$ such that

$$\|u\|_X < \rho_1, \quad \forall u \in D. \tag{3.3.2}$$

For $u \in D$ and $M(s) := \sup_{0 \le t < +\infty} \frac{|G(t,s)|}{1+t}$, by (3.3.1) and Definition 1.2.1,

$$\|Tu\|_0 = \sup_{0 \le t < +\infty} \frac{|Tu(t)|}{1+t}$$

$$\le \sup_{0 \le t < +\infty} \left( \frac{|A+Bt|}{1+t} + \frac{1}{1+t} \sum_{0 < t_k < t < +\infty} |I_{0k}(t_k, u(t_k), u'(t_k)) \right.$$

$$+ I_{1k}(t_k, u(t_k), u'(t_k))(t - t_k)| + \frac{t}{1+t} \sum_{k=1}^{+\infty} |I_{1k}(t_k, u(t_k), u'(t_k))| \right)$$

$$+ \int_0^{+\infty} \sup_{0 \le t < +\infty} \frac{|G(t,s)|}{1+t} |f(s, u(s), u'(s))| ds$$

$$\le \max\{|A|, |B|\} + \sup_{0 \le t < +\infty} \frac{1}{1+t} \left( \sum_{0 < t_k < t < +\infty} [\varphi_{k,\rho_1} + \psi_{k,\rho_1} t] \right)$$

$$+ \sup_{0 \le t < +\infty} \frac{t}{1+t} \sum_{k=1}^{+\infty} \psi_{k,\rho_1} + \int_0^{+\infty} M(s) \varphi_{\rho_1}(s) ds$$

$$\le \max\{|A|, |B|\} + \sum_{k=1}^{+\infty} \varphi_{k,\rho_1} + 2 \sum_{k=1}^{+\infty} \psi_{k,\rho_1}$$

$$+ \int_0^{+\infty} M(s) \varphi_{\rho_1}(s) ds < +\infty,$$

and

$$\|(Tu)'\|_1 = \sup_{0 \le t < +\infty} |(Tu(t))'|$$

$$\le |B| + \sum_{0 < t_k < t < +\infty} |I_{1k}(t_k, u(t_k), u'(t_k))|$$

$$+ \sum_{k=1}^{+\infty} |I_{1k}(t_k, u(t_k), u'(t_k))| + \int_t^{+\infty} |f(s, u(s), u'(s))| \, ds$$

$$\leq |B| + 2 \sum_{k=1}^{+\infty} \psi_{k,\rho_1} + \int_0^{+\infty} \varphi_{\rho_1}(s) \, ds < +\infty.$$

Therefore, $\|Tu\|_X := \max\{\|Tu\|_0, \|(Tu)'\|_1\} < +\infty$, and $TB$ is uniformly bounded in $X$.

**Step 3.** *$TD$ is equicontinuous on each finite interval* $]t_k, t_{k+1}]$, *for* $k = 0, 1, 2, \ldots$.

Consider an interval $J \subseteq ]t_k, t_{k+1}]$ and $\tau_1, \tau_2 \in J$ such that $\tau_1 \leq \tau_2$. For $u \in D$, the following limits hold uniformly as $\tau_1 \to \tau_2$:

$$\lim_{\tau_1 \to \tau_2} \left| \frac{Tu(\tau_1)}{1 + \tau_1} - \frac{Tu(\tau_2)}{1 + \tau_2} \right|$$

$$\leq \lim_{\tau_1 \to \tau_2} \left| \frac{A + B\tau_1}{1 + \tau_1} - \frac{A + B\tau_2}{1 + \tau_2} \right| + \left| \frac{1}{1 + \tau_1} \sum_{0 < t_k < \tau_1} [I_{0k}(t_k, u(t_k), u'(t_k)) \right.$$

$$+ \left. I_{1k}(t_k, u(t_k), u'(t_k))(\tau_1 - t_k)] - \frac{\tau_1}{1 + \tau_1} \sum_{k=1}^{+\infty} I_{1k}(t_k, u(t_k), u'(t_k)) \right.$$

$$- \frac{1}{1 + \tau_2} \sum_{0 < t_k < \tau_2} [I_{0k}(t_k, u(t_k), u'(t_k)) + I_{1k}(t_k, u(t_k), u'(t_k))(\tau_2 - t_k)]$$

$$+ \left. \frac{\tau_2}{1 + \tau_2} \sum_{k=1}^{+\infty} I_{1k}(t_k, u(t_k), u'(t_k)) \right| + \int_0^{+\infty} \left| \frac{G(\tau_1, s)}{1 + \tau_1} - \frac{G(\tau_2, s)}{1 + \tau_2} \right|$$

$$\times |f(s, u(s), u'(s))| ds = 0$$

and

$$\lim_{\tau_1 \to \tau_2} |(Tu)'(\tau_1) - (Tu)'(\tau_2)|$$

$$\leq \lim_{\tau_1 \to \tau_2} \left| \sum_{0 < t_k < \tau_1} I_{1k}(t_k, u(t_k), u'(t_k)) - \sum_{0 < t_k < \tau_2} I_{1k}(t_k, u(t_k), u'(t_k)) \right|$$

$$+ \left| \int_{\tau_1}^{+\infty} f(s, u(s), u'(s))\, ds - \int_{\tau_2}^{+\infty} f(s, u(s), u'(s))ds \right|$$

$$\leq \lim_{\tau_1 \to \tau_2} \sum_{\tau_1 < t_k < \tau_2} |I_{1k}(t_k, u(t_k), u'(t_k))| + \int_{\tau_1}^{\tau_2} |f(s, u(s), u'(s))|\, ds$$

$$\leq \lim_{\tau_1 \to \tau_2} \sum_{\tau_1 < t_k < \tau_2} \psi_{k,\rho_1} + \int_{\tau_1}^{\tau_2} \varphi_{\rho_1}(s)ds = 0.$$

So, $TD$ is equicontinuous on $J \subseteq ]t_k, t_{k+1}]$.

**Step 4.** *$TD$ is equiconvergent at* $t = t_i^+$, $i = 0, 1, 2, \ldots$, *and at infinity.*
In fact,

$$\left| \frac{Tu(t)}{1+t} - \lim_{t \to t_i^+} \frac{Tu(t)}{1+t} \right|$$

$$\leq \left| \frac{A + Bt}{1+t} - \frac{A + Bt_i}{1+t_i} \right| + \left| \frac{1}{1+t} \sum_{0 < t_k < t < +\infty} [I_{0k}(t_k, u(t_k), u'(t_k)) \right.$$

$$+ I_{1k}(t_k, u(t_k), u'(t_k))(t - t_k)] - \frac{t}{1+t} \sum_{k=1}^{+\infty} I_{1k}(t_k, u(t_k), u'(t_k))$$

$$- \frac{1}{1+t_i} \sum_{0 < t_k < t_i^+} [I_{0k}(t_k, u(t_k), u'(t_k)) + I_{1k}(t_k, u(t_k), u'(t_k))(t_i - t_k)]$$

$$\left. + \frac{t_i}{1+t_i} \sum_{k=1}^{+\infty} I_{1k}(t_k, u(t_k), u'(t_k)) \right| + \int_0^{+\infty} \left| \frac{G(t,s)}{1+t} - \frac{G(t_i, s)}{1+t_i} \right|$$

$$\times \varphi_{\rho_1}(s)ds \longrightarrow 0, \text{ uniformly as } t \to t_i^+,$$

and

$$\left| (Tu)'(t) - \lim_{t \to t_i^+} (Tu)'(t) \right|$$

$$\leq \left| \sum_{0 < t_k < t < +\infty} I_{1k}(t_k, u(t_k), u'(t_k)) - \sum_{0 < t_k < t_i^+} I_{1k}(t_k, u(t_k), u'(t_k)) \right|$$

$$+ \left| \int_{t_i}^{+\infty} |f(s, u(s), u'(s))| - \int_{t}^{+\infty} |f(s, u(s), u'(s))| \, ds \right|$$

$$\leq \left| \sum_{0 < t_k < t < +\infty} I_{1k}(t_k, u(t_k), u'(t_k)) - \sum_{0 < t_k < t_i^+} I_{1k}(t_k, u(t_k), u'(t_k)) \right|$$

$$+ \int_{t_i}^{t} \varphi_{\rho_1}(s) ds \longrightarrow 0, \text{ uniformly as } t \to t_i^+.$$

Therefore, $TD$ is equiconvergent at $t = t_i^+$, $i = 0, 1, 2, \ldots$.

Moreover, as $\frac{G(t,s)}{1+t}$ is bounded in $[0, +\infty[$ and $f$ is bounded on $D$ by an $L^1$-function, by Lebesgue's Dominated Convergence Theorem, we have

$$\left| \frac{Tu(t)}{1+t} - \lim_{t \to +\infty} \frac{Tu(t)}{1+t} \right|$$

$$= \left| \left( \frac{A + Bt}{1+t} + \frac{1}{1+t} \sum_{0 < t_k < t < +\infty} [I_{0k}(t_k, u(t_k), u'(t_k)) \right. \right.$$

$$+ \left. I_{1k}(t_k, u(t_k), u'(t_k))(t - t_k)] - \frac{t}{1+t} \sum_{k=1}^{+\infty} I_{1k}(t_k, u(t_k), u'(t_k)) \right.$$

$$+ \left. \frac{1}{1+t} \int_{0}^{+\infty} G(t,s) f(s, u(s), u'(s)) ds \right) - \left( B + \lim_{t \to +\infty} \frac{1}{1+t} \right.$$

$$\times \sum_{0 < t_k < t < +\infty} [I_{0k}(t_k, u(t_k), u'(t_k)) + I_{1k}(t_k, u(t_k), u'(t_k))(t - t_k)]$$

$$\left. - \sum_{k=1}^{+\infty} I_{1k}(t_k, u(t_k), u'(t_k)) + \int_{0}^{+\infty} \lim_{t \to +\infty} \frac{G(t,s)}{1+t} f(s, u(s), u'(s)) ds \right) \right|$$

$$\leq \left| \frac{A + Bt}{1+t} - B \right| + \left| \frac{1}{1+t} \sum_{0 < t_k < t < +\infty} [I_{0k}(t_k, u(t_k), u'(t_k)) \right.$$

$$+ \left. I_{1k}(t_k, u(t_k), u'(t_k))(t - t_k)] - \sum_{0 < t_k < t < +\infty} I_{1k}(t_k, u(t_k), u'(t_k)) \right|$$

$$+ \left| \sum_{k=1}^{+\infty} I_{1k}(t_k, u(t_k), u'(t_k)) - \frac{t}{1+t} \sum_{k=1}^{+\infty} I_{1k}(t_k, u(t_k), u'(t_k)) \right|$$

$$+ \int_0^{+\infty} \left| \frac{G(t,s)}{1+t} - \lim_{t \to +\infty} \frac{G(t,s)}{1+t} \right| \varphi_{\rho_1}(s) ds.$$

As each modulus tends to 0, uniformly on $u \in D$, as $t \to +\infty$, then

$$\left| \frac{Tu(t)}{1+t} - \lim_{t \to +\infty} \frac{Tu(t)}{1+t} \right| \longrightarrow 0, \text{ uniformly on } u \in D, \text{ as } t \to +\infty.$$

By similar arguments,

$$\left| (Tu)'(t) - \lim_{t \to +\infty} (Tu)'(t) \right|$$

$$= \left| \left( B + \sum_{0 < t_k < t < +\infty} I_{1k}(t_k, u(t_k), u'(t_k)) - \sum_{k=1}^{+\infty} I_{1k}(t_k, u(t_k), u'(t_k)) \right. \right.$$

$$\left. - \int_t^{+\infty} f(s, u(s), u'(s)) ds \right) - \left( B + \lim_{t \to +\infty} \int_t^{+\infty} f(s, u(s), u'(s)) ds \right) \right|$$

$$\leq \left| \sum_{0 < t_k < t < +\infty} I_{1k}(t_k, u(t_k), u'(t_k)) - \sum_{k=1}^{+\infty} I_{1k}(t_k, u(t_k), u'(t_k)) \right|$$

$$+ \int_t^{+\infty} |f(s, u(s), u'(s))| \, ds$$

$$\leq \left| \sum_{0 < t_k < t < +\infty} I_{1k}(t_k, u(t_k), u'(t_k)) - \sum_{k=1}^{+\infty} I_{1k}(t_k, u(t_k), u'(t_k)) \right|$$

$$+ \int_t^{+\infty} \varphi_{\rho_1}(s) ds \longrightarrow 0, \text{ uniformly on } u \in D \text{ as } t \to +\infty,$$

that is, $TD$ is equiconvergent at $+\infty$.

So, by Lemma 3.2.3 adapted to the impulsive case, $TD$ is relatively compact and $T$ is completely continuous.

**Step 5.** $T\Omega \subset \Omega$, *for $\Omega$ is a nonempty, bounded, closed and convex subset of $X$.*

Consider a subset $\Omega \subset X$ defined as $\Omega := \{u \in X : \|u\|_X \leq \rho_2\}$ with

$$\rho_2 := \max \left\{ \rho_1, \max\{|A|, |B|\} + \sum_{k=1}^{+\infty} (\varphi_{k,\rho_1} + 2\psi_{k,\rho_1}) \right.$$
$$\left. + \int_0^{+\infty} M_1(s)\varphi_{\rho_1}(s)ds \right\},$$

where $\rho_1$ is given by (3.3.2) and

$$M_1(s) := \max \left\{ 1, \sup_{0 \leq t < +\infty} \frac{|G(t,s)|}{1+t} \right\}.$$

Remark that $0 \leq M_1(s) \leq 1$, for $s \in [0, +\infty[$, and, therefore,

$$\int_0^{+\infty} M_1(s)\,\varphi_{\rho_1}(s)ds \leq \int_0^{+\infty} \varphi_{\rho_1}(s)ds < +\infty.$$

For $u \in \Omega$,

$$\|Tu\|_0 = \sup_{0 \leq t < +\infty} \frac{|Tu(t)|}{1+t}$$

$$\leq \sup_{0 \leq t < +\infty} \frac{|A + Bt|}{1+t} + \frac{1}{1+t} \sum_{0 < t_k < t < +\infty} |I_{0k}(t_k, u(t_k), u'(t_k))$$

$$+ I_{1k}(t_k, u(t_k), u'(t_k))(t - t_k)| + \frac{t}{1+t} \sum_{k=1}^{+\infty} |I_{1k}(t_k, u(t_k), u'(t_k))|$$

$$+ \int_0^{+\infty} \sup_{0 \leq t < +\infty} \frac{|G(t,s)|}{1+t} |f(s, u(s), u'(s))|\, ds$$

$$\leq \max\{|A|, |B|\} + \sup_{0 \leq t < +\infty} \frac{1}{1+t} \left( \sum_{k=1}^{+\infty} \varphi_{k,\rho_1} + \psi_{k,\rho_1} (t - t_k) \right)$$

$$+ \sup_{0 \leq t < +\infty} \frac{t}{1+t} \sum_{k=1}^{+\infty} \psi_{k,\rho_1} + \int_0^{+\infty} M_1(s)\,\varphi_{\rho_1}(s)ds$$

$$\leq \max\{|A|, |B|\} + \sum_{k=1}^{+\infty} \varphi_{k,\rho_1} + 2\sum_{k=1}^{+\infty} \psi_{k,\rho_1} + \int_0^{+\infty} M_1(s)\,\varphi_{\rho_1}(s)ds$$

$$< \rho_2,$$

and

$$\|(Tu)'\|_1 = \sup_{0 \le t < +\infty} \left|(Tu(t))'\right|$$

$$\le |B| + \sum_{0 < t_k < t < +\infty} |I_{1k}(t_k, u(t_k), u'(t_k))|$$

$$+ \sum_{k=1}^{+\infty} |I_{1k}(t_k, u(t_k), u'(t_k))| + \int_t^{+\infty} |f(s, u(s), u'(s))| \, ds$$

$$\le |B| + 2 \sum_{k=1}^{+\infty} \psi_{k, \rho_1} + \int_0^{+\infty} \varphi_{\rho_1}(s) ds < \rho_2.$$

Therefore, $\|Tu\|_X < \rho_2$ and $T\Omega \subset \Omega$.

Then by Schauder's fixed-point theorem, $T$ has at least one fixed point $u \in X$. So, the problem (9.1.1)–(3.1.3) has a solution $u \in X$.

Moreover, $u$ is bounded if $B = 0$, and unbounded if $B \neq 0$.     □

## 3.4. Example

Consider the second-order two-point impulsive problem composed by the fully differential equation in the half-line

$$u''(t) = \frac{(1 + e^{-t}) u(t) + (u'(t))^3}{1 + t^4}, \quad \text{a.e. } t > 0, \tag{3.4.1}$$

the boundary conditions

$$u(0) = 1, \quad u'(+\infty) = \frac{1}{2}, \tag{3.4.2}$$

and the impulsive effects

$$\Delta u(k) = \frac{1}{k^3} \frac{1}{\left[(u(k))^2 + 1\right]\left[(u'(k))^2 + 1\right]}, \tag{3.4.3}$$

$$\Delta u'(k) = \frac{|u(k)| + |u'(k)|}{k^\alpha},$$

with $k = 1, 2, 3, \ldots, \alpha \in \mathbb{R}, \alpha > 2$.

We point out that problem (3.4.1)–(3.4.3) is a particular case of (3.1.1)–(3.1.3) with

$$f(t, x, y) = \frac{(1 + e^{-t})x + y^3}{1 + t^4},$$

$$A = 1, \quad B = \frac{1}{2},$$

$$t_k = k, \quad k \in \mathbb{N},$$

$$I_{0k}(t_k, x, y) = \frac{1}{k^2} \frac{1}{(x^2 + 1)(y^2 + 1)},$$

$$I_{1k}(t_k, x, y) = \frac{|x| + |y|}{k^\alpha}.$$

As $f$ is an $L^1$-Carathéodory function in $[0, +\infty[$, with

$$\varphi_\rho(t) := \frac{2\rho(1 + t) + \rho^3}{1 + t^4},$$

for $|x| < \rho(1 + t)$ and $|y| < \rho$, the function

$$I_{0k}(k, x, y) := \frac{1}{k^3 (x^2 + 1) (y^2 + 1)}$$

is a Carathéodory sequence for every $x, y \in \mathbb{R}$, with $\Psi^0_{k,\rho} := \frac{1}{k^3}$, and

$$I_{1k}(k, x, y) := \frac{|x| + |y|}{k^\alpha}$$

is also a Carathéodory sequence for $\rho > 0$ such that $|x| < \rho(1 + k)$ and $|y| < \rho$, with $\Psi^1_{k,\rho} := \frac{\rho(k+2)}{k^\alpha}$ $(\alpha > 2)$.

Therefore, by Theorem 3.3.1, problem (3.4.1)–(3.4.3) has at least an unbounded solution.

# Part II

# Homoclinic Solutions
# and Lidstone Problems

# Introduction

Qualitative analysis of differential equations has had an increasingly important role, especially the analytic study of their asymptotic behavior and stability.

A *homoclinic* orbit is a trajectory of a flow of a dynamical system which joins a saddle equilibrium point to itself. If a path in the phase space of a dynamical system joins two different equilibrium points, it receives the name of a *heteroclinic* orbit.

Homoclinic trajectory, heteroclinic connection and heteroclinic cycle

The interest in these trajectories goes far beyond mathematics itself as homoclinic and heteroclinic solutions appear in a variety of mathematical models born in mechanics, chemistry, or biology.

The history of these homoclinic and heteroclinic solutions is already long. In addition to the phase portrait analysis, whose applicability is restricted to autonomous differential equations of second order, the study of these solutions started with a geometric approach. Poincaré, Melnikov, and Smale were some of the first names to cover this topic in the nineteenth century. At the end of the last century, a more functional and analytical approach gave new tools like variational methods and the theory of critical points. It is worth highlighting Ambrosetti, Ekeland and Rabinowitz (see [44] and references therein).

This part is separated into three chapters, and each one provides the existence of homoclinic solutions for higher order nonlinear BVPs, not necessarily autonomous.

Chapter 4 will be addressed to problems with second-order equations. Three different applications will be presented to illustrate the main results of the chapter: a problem with discontinuity in time, an application to a Duffing equation, and another over a forced cantilever beam equation with damping.

Chapter 5 confirms the existence of homoclinic solutions to some fourth-order BVPs. A generic example and an application to a Bernoulli–Euler–v. Karman BVP complete the chapter.

Finally, Chapter 6 focuses the attention on Lidstone's BVPs, putting a link between the solutions of Lidstone BVPs in the whole real line and homoclinic solutions. The results of this chapter will be applied to an infinite beam resting on granular foundations with moving loads.

# Chapter 4

# Homoclinic Solutions
# for Second-Order Problems

## 4.1. Introduction

The existence of homoclinic solutions for autonomous and nonautonomous differential equations and Hamiltonian systems is an important subject in qualitative theory. It can be considered as a special case of the so-called *convergent solutions*, i.e., solutions defined on the half-line (or the real line), and having a finite limit to $+\infty$ (respectively $\pm\infty$), see [16].

In this chapter, we consider the second-order discontinuous equation in the real line,

$$u''(t) - ku(t) = f(t, u(t), u'(t)), \quad \text{a.e.} \, t \in \mathbb{R}, \tag{4.1.1}$$

with $k > 0$ and $f : \mathbb{R}^3 \to \mathbb{R}$ an $L^1$-Carathéodory function. The main purpose is to find homoclinic orbits to $0$, that is, nontrivial solutions of (4.1.1) such that

$$u(\pm\infty) := \lim_{t \to \pm\infty} u(t) = 0, u'(\pm\infty) := \lim_{t \to \pm\infty} u'(t) = 0. \tag{4.1.2}$$

Several works prove the existence of homoclinic and heteroclinic solutions for small perturbations (see [48, 156]), or deal with some superquadratic or subquadratic conditions at infinity (see [135, 140]) or with asymptotically quadratic conditions (see [55]). Another point of view is to obtain a homoclinic orbit as a limit of $2kT$-periodic solutions of a certain sequence of periodic boundary value problems (see [10, 74, 85]). The main arguments used in this method apply variational methods, upper and lower solutions and fixed point theory (see [17, 28, 138, 146]).

Equation (4.1.1) arises in several real phenomena, for instance, as the study of traveling wave fronts for parabolic reaction–diffusion equations with a local reaction term, and generalizes several classical equations such as Duffing-type equations (see [76, 130]) or Liénard-like systems (see [154]).

In this chapter, we combine the method of lower and upper solutions, not necessarily ordered, as suggested in [75, 113]. Moreover, our result improves the literature as the existence and localization of homoclinic solutions is proved without extra assumptions on the growth, sign or asymptotic behavior of the nonlinear part.

## 4.2. Preliminaries

Define the space

$$X_{H2} = \left\{ x \in C^1(\mathbb{R}) : \lim_{|t| \to +\infty} x(t) \in \mathbb{R} \right\},$$

with the norm $\|x\|_{X_{H2}} = \max\{\|x\|_\infty, \|x'\|_\infty\}$, where $\|y\|_\infty := \sup_{t \in \mathbb{R}} |y(t)|$. In this way, $(X_{H2}, \| \cdot \|_{X_{H2}})$ is a Banach space (see [149, 153]).

An important property of functions on space $X_{H2}$ is shown in the following lemma.

**Lemma 4.2.1.** *Let $x \in C^n(\mathbb{R})$, $n \in \mathbb{N}$, $n \geq 1$. If $x(+\infty) = l \in \mathbb{R}$ then $x^{(n)}(+\infty) = 0$, for $n \geq 1$.*

**Proof.** In the case where $x(+\infty) = l$, for any $\delta_0 > 0$, there exists $T_0 > 0$ such that for $t > T_0$, one has $|x(t) - l| < \delta_0$.

For $n = 1$, take $h > 0, \delta_0 = \frac{h \delta_1}{2}$ and $t > T_1$, for some $T_1 > 0$. Therefore, for $t > \max\{T_0, T_1\}$, one has

$$|x'(t)| = \lim_{h \to 0} \frac{|x(t+h) - x(t)|}{h} = \lim_{h \to 0} \frac{|x(t+h) - l + l - x(t)|}{h}$$

$$\leq \lim_{h \to 0} \frac{|x(t+h) - l| + |x(t) - l|}{h} \leq \lim_{h \to 0} \frac{\frac{h \delta_1}{2} + \frac{h \delta_1}{2}}{h} = \delta_1,$$

for any $\delta_1 > 0$, that is, $x'(+\infty) = 0$.

For $n > 1$, the proof follows by the mathematical induction.

The case $x(-\infty) = l$ can be proved by using the same technique. □

The following result will play an important role in the proof of the main result, giving a solution of some linear second-order problem via Green's functions.

**Lemma 4.2.2 ([3]).** *If $h \in L^1(\mathbb{R})$, then problem*

$$\begin{cases} u''(t) - ku(t) = h(t), & \text{a.e. } t \in \mathbb{R}, \\ u(\pm\infty) = u'(\pm\infty) = 0 \end{cases} \tag{4.2.1}$$

*has a unique solution in $X_{H2}$. Moreover, this solution can be expressed as*

$$u(t) = \int_{-\infty}^{+\infty} G(t,s)h(s)ds, \tag{4.2.2}$$

*where*

$$G(t,s) = -\frac{1}{2\sqrt{k}}e^{-\sqrt{k}|s-t|}. \tag{4.2.3}$$

**Proof.** The homogeneous solution of the linear equation is given by

$$u(t) = c_1 e^{\sqrt{k}t} + c_2 e^{-\sqrt{k}t}, \quad \text{for } c_1, c_2 \in \mathbb{R}.$$

As the null function is the only solution of the homogeneous problem associated to (4.2.1), its solution is given by

$$u(t) = -\frac{1}{2\sqrt{k}} \int_{-\infty}^{+\infty} e^{-\sqrt{k}|s-t|} h(s)ds.$$

For $G(t,s) := -\frac{1}{2}e^{-\sqrt{k}|s-t|}$, one has

$$u(t) = \int_{-\infty}^{+\infty} G(t,s)h(s)ds.$$

$\square$

Some trivial properties can easily be proved for Green's functions.

**Remark 4.2.3.** The above Green's functions verify the following properties:

- $G(t,s)$ and $\frac{\partial G(t,s)}{\partial t}$ are continuous,

- $\lim_{|t|\to+\infty} G(t,s) = 0$,

- $\lim_{|t|\to+\infty} \frac{\partial G(t,s)}{\partial t} = 0$.

To deal with the lack of compactness of set $X_{H2}$, next compactness criterion plays a key role, following arguments suggested in [51, 128, 149].

**Theorem 4.2.4.** *A set $M \subset X_{H2}$ is compact if the following conditions hold:*

(i) *both $\{t \to x(t) : x \in M\}$ and $\{t \to x'(t) : x \in M\}$ are uniformly bounded;*

(ii) *both $\{t \to x(t) : x \in M\}$ and $\{t \to x'(t) : x \in M\}$ are equicontinuous in any compact interval of $\mathbb{R}$;*

(iii) *both $\{t \to x(t) : x \in M\}$ and $\{t \to x'(t) : x \in M\}$ are equiconvergent at $\pm\infty$, that is, given $\epsilon > 0$, there exists $T(\epsilon) > 0$ such that $|f(t) - f(\pm\infty)| < \epsilon$ and $|f'(t) - f'(\pm\infty)| < \epsilon$ for all $|t| > T(\epsilon)$ and $f \in M$.*

**Proof.** In order to prove that the subset $M$ is relatively compact in $X_{H2}$, as we are in a Banach space, we only need to show that $M$ is totally compact or bounded in $X_{H2}$, that is, for $\epsilon > 0$, $M$ has a finite $\epsilon$-net.

For any given $\epsilon > 0$, by (i)–(iii), there exist constants $A > 0, \delta > 0$, and an integer $N > 0$, such that

- $|x(t_1) - x(t_2)| \leq \frac{\epsilon}{3}, |x'(t_1) - x'(t_2)| \leq \frac{\epsilon}{3}$ with $t_1, t_2 < -N$ or $t_1, t_2 > N$ and $x \in M, \|x\|_{X_{H2}} \leq A$;
- $|x(t_1) - x(t_2)| \leq \frac{\epsilon}{3}, |x'(t_1) - x'(t_2)| \leq \frac{\epsilon}{3}$ with $t_1, t_2 \in [-N, N]$ and $|t_1 - t_2| < \delta, x \in X_{H2}$.

Define $X_{[-N,N]} = \{x|_{[-N,N]} : x \in X_{H2}\}$. For $x \in X_{[-N,N]}$, define

$$\|x\|_N = \max\left\{ \sup_{t \in [-N,N]} |x(t)|, \sup_{t \in [-N,N]} |x'(t)| \right\}.$$

It can be proved that $X_{[-N,N]}$ is a Banach space with the norm $\|\cdot\|_N$.

Let $M_{[-N,N]} = \{t \to x(t), t \in [-N, N] : x \in M\}$. Then $M_{[-N,N]}$ is a subset of $X_{[-N,N]}$. By the Arzèla–Ascoli theorem, $M_{[-N,N]}$ is relatively compact in $X_{[-N,N]}$. Thus, there exist $x_1, x_2, \ldots, x_k \in M$ such that $\|x - x_i\|_N \leq \frac{\epsilon}{3}$, for any $x \in M$ and $i = 1, 2, \ldots, k$.

Therefore, for $x \in M$, we find that for $j = 0, 1$,

$$\|x^{(j)} - x_i^{(j)}\|_X = \max\left\{ \sup_{t \in \mathbb{R}} |x^{(j)}(t) - x_i^{(j)}(t)| \right\}$$

$$= \max\left\{ \begin{array}{l} \sup_{t \leq -N} |x^{(j)}(t) - x_i^{(j)}(t)| \\ \sup_{|t| \leq N} |x^{(j)}(t) - x_i^{(j)}(t)| \\ \sup_{t \geq N} |x^{(j)}(t) - x_i^{(j)}(t)| \end{array} \right\}$$

$$\leq \max\left\{ \sup_{t \leq -N} |x^{(j)}(t) - x_i^{(j)}(t)|, \frac{\epsilon}{3}, \sup_{t \geq N} |x^{(j)}(t) - x_i^{(j)}(t)| \right\}.$$

Moreover,

$$\sup_{t \leq -N} |x(t) - x_i(t)| \leq \sup_{t \leq -N} |x(t) - x(-N)| + |x(-N) - x_i(-N)|$$

$$+ \sup_{t \leq -N} |x_i(-N) - x_i(t)| \leq \frac{\epsilon}{3} + \frac{\epsilon}{3} + \frac{\epsilon}{3} = \epsilon.$$

Similarly, we can prove that all $\sup_{|t|>N} |x^{(j)}(t) - x_i^{(j)}(t)| \leq \epsilon$.

So, for any $\epsilon > 0$, $M$ has a finite $\epsilon$-net $\{U_{x_1}, U_{x_2}, \ldots, U_{x_k}\}$, that is, $M$ is totally bounded in $X_{H2}$. Hence, $M$ is relatively compact in $X_{H2}$. $\square$

To provide the localization part of the main result, lower and upper solutions technique is used, based on the following definition.

**Definition 4.2.5.** A function $\alpha \in X_{H2}$ is said to be a lower solution of problem (4.1.1),(4.1.2) if

$$\alpha''(t) - k\,\alpha(t) \geq f(t, \alpha(t), \alpha'(t)), \quad \text{a.e. } t \in \mathbb{R}, \quad \text{and} \quad \alpha(\pm\infty) \leq 0.$$

A function $\beta \in X_{H2}$ is an upper solution if the reversed inequalities hold.

Usually, in the literature, these functions have some order relation: well ordered or reversed ordered. However, next definition can be applied to $\alpha(t)$ and $\beta(t)$ with no definite order.

**Definition 4.2.6.** Functions $\alpha, \beta \in X_{H2}$ are a pair of lower and upper solutions of problem (4.1.1),(4.1.2), respectively, if

$$\begin{cases} \alpha''(t) - k\,\overline{\alpha}(t) \geq f(t, \overline{\alpha}(t), \alpha'(t)), & t \in \mathbb{R}, \\ \beta''(t) - k\,\beta(t) \leq f(t, \beta(t), \beta'(t)), & t \in \mathbb{R}, \\ \overline{\alpha}(\pm\infty) \leq 0, \quad \beta(\pm\infty) \geq 0, \end{cases}$$

where $\overline{\alpha}(t) = \alpha(t) - \sup_{t \in \mathbb{R}} |\alpha(t) - \beta(t)|$.

## 4.3. Existence and localization of homoclinics

First result requires that lower and upper solutions are well ordered to guarantee the existence of homoclinic solutions of problem (4.1.1),(4.1.2).

**Theorem 4.3.1.** *Let* $f : \mathbb{R}^3 \to \mathbb{R}$ *be an* $L^1$-*Carathéodory function and* $\alpha, \beta \in X_{H2}$ *be lower and upper solutions of problem* (4.1.1),(4.1.2), *respectively, with*

$$\alpha(t) \le \beta(t), \quad \forall t \in \mathbb{R}. \tag{4.3.1}$$

*If $f(t,x,y)$ is monotone in $y$ (nonincreasing or nondecreasing) for $(t,x) \in \mathbb{R}^2$ fixed, then problem (4.1.1),(4.1.2) has a homoclinic solution $u \in X_{H2}$ such that $\alpha(t) \le u(t) \le \beta(t), \forall t \in \mathbb{R}$.*

**Proof.** Consider the modified equation

$$u''(t) - ku(t) = f(t, \delta(t, u(t)), u'(t)), \quad \text{a.e. } t \in \mathbb{R}, \tag{4.3.2}$$

where function $\delta : \mathbb{R}^2 \to \mathbb{R}$ is given by

$$\delta(t, u(t)) = \begin{cases} \beta(t), & u(t) > \beta(t), \\ u(t), & \alpha(t) \le u(t) \le \beta(t), \\ \alpha(t), & u(t) < \alpha(t). \end{cases}$$

**Step 1.** *The modified problem (4.3.2),(4.1.2) has a solution.*
Define the operator $T : X_{H2} \to X_{H2}$ by

$$Tu(t) = \int_{-\infty}^{+\infty} G(t,s) F_u(s) ds,$$

where

$$F_u(t) = f(t, \delta(t, u(t)), u'(t)),$$

and $G(t,s)$ is the Green Function given by Lemma 4.2.2. So, it is enough to prove that $T$ has a fixed point, which is done in the following claims.

**Claim 1.1.** $T : X_{H2} \to X_{H2}$ *is well defined.*
Let $u \in X_{H2}$. If $f$ is an $L^1$-Carathéodory function, then $Tu$ is continuous. For $r_0 > 0$ such that

$$r_0 > \max\{\|\alpha\|_\infty, \|\beta\|_\infty\}, \tag{4.3.3}$$

there exists $\varphi_{r_0}$ with $|f(t,x,y)| \le \varphi_{r_0}(t)$, for $\sup_{t \in \mathbb{R}}\{|x(t)|, |y(t)|\} < r_0$ and a.e. $t \in \mathbb{R}$. As $Tu$ and $(Tu)'$ are continuous, passing to the limit, by the Lebesgue Dominated Theorem and Remark 4.2.3,

$$\lim_{|t|\to\infty} (Tu)(t) = \int_{-\infty}^{+\infty} \lim_{|t|\to\infty} G(t,s) F_u(s) ds = 0,$$

$$\lim_{|t|\to\infty} (Tu)'(t) = \int_{-\infty}^{+\infty} \lim_{|t|\to\infty} \frac{\partial G(t,s)}{\partial t} F_u(s) ds = 0,$$

and, therefore, $Tu \in X_{H2}$.

**Claim 1.2.** *T is compact.*

Let

$$M(s) := \max \left\{ \sup_{t \in \mathbb{R}} |G(t,s)|, \sup_{t \in \mathbb{R}} \left| \frac{\partial G(t,s)}{\partial t} \right| \right\}.$$

Consider a bounded set $B \subset X_{H2}$ defined by

$$B := \{ u \in X_{H2} : \|u\|_{X_{H2}} < r_1 \},$$

for some $r_1 > 0$, such that $r_1 > \max \left\{ r_0, \int_{-\infty}^{+\infty} M(s) \varphi_{r_0}(s) ds \right\}$ with $r_0$ given by (4.3.3). Then, for $t \in \mathbb{R}$,

$$|Tu(t)| \le \int_{-\infty}^{+\infty} M(s) |F_u(s)| ds \le \int_{-\infty}^{+\infty} M(s) \varphi_r(s) ds < r_1,$$

and analogously $|(Tu)'(t)| < r_1$. Therefore, $TB$ is bounded and $TB \subset B$.

For $a > 0$ and $t_1, t_2 \in [-a, a]$, because of the continuity of the Green's functions and its derivative, one has

$$\lim_{t_1 \to t_2} |Tu(t_1) - Tu(t_2)|$$

$$\le \int_{-\infty}^{+\infty} \lim_{t_1 \to t_2} |G(t_1, s) - G(t_2, s)| |F_u(s)| ds = 0,$$

$$\lim_{t_1 \to t_2} |(Tu)'(t_1) - (Tu)'(t_2)|$$

$$\le \int_{-\infty}^{+\infty} \lim_{t_1 \to t_2} \left| \frac{\partial G}{\partial t}(t_1, s) - \frac{\partial G}{\partial t}(t_2, s) \right| |F_u(s)| ds = 0.$$

So, $TB$ is equicontinuous.

To prove that $TB$ is equiconvergent at $\pm\infty$, note that

$$\left| Tu(t) - \lim_{t \to \pm\infty} (Tu(t)) \right| \le \int_{-\infty}^{+\infty} |G(t,s)| |F_u(s)| ds$$

$$\le \int_{-\infty}^{+\infty} |G(t,s)| \varphi_r(s) ds \longrightarrow 0, t \to \pm\infty,$$

$$\left| (Tu)'(t) - \lim_{t \to \pm\infty} (Tu)'(t) \right| \le \int_{-\infty}^{+\infty} \left| \frac{\partial G}{\partial t}(t,s) \right| |F_u(s)| ds$$

$$\le \int_{-\infty}^{+\infty} \left| \frac{\partial G}{\partial t}(t,s) \right| \varphi_r(s) ds \longrightarrow 0, t \to \pm\infty.$$

Therefore, by Theorem 4.2.4, $T$ is compact, and by Theorem 1.2.6, $T$ has at least one fixed point $u \in X_{H2}$.

**Step 2.** *Every solution of the modified problem* (4.3.2),(4.1.2) *is a solution of the initial problem* (4.1.1),(4.1.2).

Let $u$ be a solution of problem (4.3.2),(4.1.2). In order to obtain this step, it is sufficient to prove that

$$\alpha(t) \le u(t) \le \beta(t), \quad \forall t \in \mathbb{R}.$$

Suppose, by contradiction, that there exists $t \in \mathbb{R}$ such that $\alpha(t) > u(t)$ and define

$$\inf_{t \in \mathbb{R}} (u(t) - \alpha(t)) < 0.$$

This infimum cannot be attained at $\pm\infty$. Otherwise, by (4.1.2) and Definition 4.2.5, this contradiction holds:

$$0 > u(\pm\infty) - \alpha(\pm\infty) \ge 0.$$

So, there is $t_* \in \mathbb{R}$ such that

$$\min_{t \in \mathbb{R}} (u(t) - \alpha(t)) = u(t_*) - \alpha(t_*) < 0.$$

Then there exists an interval $[t_-, t_+]$ such that $t_* \in [t_-, t_+]$ and $u(t) - \alpha(t) < 0$, $u''(t) - \alpha''(t) \ge 0$ a.e. $t \in [t_-, t_+]$. Also, $u'(t) - \alpha'(t) \le 0$, for $t \in [t_-, t_*]$, and $u'(t) - \alpha'(t) \ge 0$ for $t \in [t_*, t_+]$.

If $f(t, x, y)$ is nonincreasing in $y$, for $t \in [t_*, t_+]$ then the following contradiction is achieved

$$0 \le \int_{t_*}^{t} u''(s) - \alpha''(s)ds = \int_{t_*}^{t} [f(s, \delta(s, u(s)), u'(s)) + ku(s) - \alpha''(s)]ds$$

$$\le \int_{t_*}^{t} [f(s, \alpha(s), \alpha'(s)) + ku(s) - \alpha''(s)]ds$$

$$\le k \int_{t_*}^{t} u(s) - \alpha(s)ds < 0.$$

By the previous arguments, a similar contradiction holds if $f$ is nondecreasing, but with an integration on $[t_-, t_*] \subset [t_-, t_+]$.

So, $\alpha(t) \le u(t), \forall t \in \mathbb{R}$. In a similar way, it can be proved that $\beta(t) \ge u(t), \forall t \in \mathbb{R}$.   □

If the nonlinearity $f$ verifies an anti-symmetric-type property, there is also homoclinic solutions for the symmetric equation

$$-u''(t) + ku(t) = f(t, u(t), u'(t)), \quad t \in \mathbb{R}. \tag{4.3.4}$$

**Theorem 4.3.2.** *Let $\alpha, \beta \in X_{H2}$ be lower and upper solutions of problem (4.1.1),(4.1.2), respectively, verifying (4.3.1). If $f : \mathbb{R}^3 \to \mathbb{R}$ is an $L^1$-Carathéodory function, with $f(t, x, y)$ monotone in $y$, for $(t, x) \in \mathbb{R}^2$ fixed, satisfying*

$$f(t, -x, -y) = -f(t, x, y), \quad \forall (t, x, y) \in \mathbb{R}^3, \tag{4.3.5}$$

*then there is a pair of homoclinic solutions $(u, -u) \in X_{H2}^2$ such that $u$ is a solution of problem (4.1.1),(4.1.2) and $-u$ solution of (4.3.4), (4.1.2), verifying*

$$\alpha(t) \le u(t) \le \beta(t),$$

$$-\beta(t) \le -u(t) \le -\alpha(t), \quad \forall t \in \mathbb{R}.$$

**Proof.** Let $\alpha \in X_{H2}$ be lower and upper solutions of problem (4.1.1),(4.1.2). Then, by (4.3.5),

$$-\alpha''(t) + k\,\alpha(t) = -\left[\alpha''(t) - k\,\alpha(t)\right]$$

$$\le -f(t, \alpha(t), \alpha'(t)) = f(t, -\alpha(t), -\alpha'(t)), \quad \text{for } t \in \mathbb{R},$$

that is, $-\alpha(t)$ is an upper solution of problem (4.3.4),(4.1.2).

Analogously, it can be proved that $-\beta(t)$ is a lower solution of problem (4.3.4),(4.1.2).

So, by Theorem 4.3.1, there is a solution $-u$ of problem (4.3.4),(4.1.2), such that

$$-\beta(t) \le -u(t) \le -\alpha(t), \quad \forall t \in \mathbb{R}. \qquad \square$$

The well-ordered relation (4.3.1) can be removed if lower and upper functions are defined as a pair of functions, applying a translation technique suggested in [63].

In this case, the main theorem can be formulated in the following way.

**Theorem 4.3.3.** *Let $f : \mathbb{R}^3 \to \mathbb{R}$ be an $L^1$-Carathéodory function and $\alpha, \beta \in X_{H2}$ a pair of lower and upper solutions of problem (4.1.1),(4.1.2), respectively, according to Definition 4.2.6.*

*If $f(t, x, y)$ is monotone in $y$ (nonincreasing or nondecreasing) for $(t, x) \in \mathbb{R}^2$ fixed, then problem (4.1.1), (4.1.2) has a homoclinic solution $u \in X_{H2}$ such that $\overline{\alpha}(t) \le u(t) \le \beta(t)$.*

The proof is similar to Theorem 4.3.1 replacing the truncature function $\delta$ by $\overline{\delta} : \mathbb{R}^2 \to \mathbb{R}$ given as

$$\overline{\delta}(t, u(t)) = \begin{cases} \beta(t), & u(t) > \beta(t), \\ u(t), & \overline{\alpha}(t) \le u(t) \le \beta(t), \\ \overline{\alpha}(t), & u(t) < \overline{\alpha}(t). \end{cases}$$

Note that $\alpha$ and $\beta$ do not need to be well ordered or even ordered at all.

## 4.4.   Example of a discontinuous BVP

Consider the second-order, nonlinear and discontinuous BVP

$$\begin{cases} u''(t) - u(t) = \dfrac{\mathrm{sgn}(t)u^3(t) + 0.1 - 100u'(t)}{1 + t^2}, & t \in \mathbb{R}, \\ u(\pm\infty) = u'(\pm\infty) = 0. \end{cases} \tag{4.4.1}$$

where

$$\mathrm{sgn}(t) = \begin{cases} 1, & t \ge 0, \\ -1, & t < 0. \end{cases}$$

The nonlinear and discontinuous function $f : \mathbb{R}^3 \to \mathbb{R}$ defined by

$$f(t, x, y) := \frac{\mathrm{sgn}(t)x^3 + 0.1 - 100y}{1 + t^2}$$

is monotone in $y$ for $(t, x) \in \mathbb{R}^2$ fixed and for $|x|, |y| < \rho$, and an $L^1$-Carathéodory function with $\varphi_\rho(t) = \frac{\rho^3 + 0.1 + 100\rho}{1 + t^2}$.

Functions $\alpha(t) = \arctan(t)$ and $\beta(t) \equiv 0$ are, respectively, a pair of lower and upper solutions of problem (4.4.1) according to Definition 4.2.6 with $\overline{\alpha}(t) = \arctan(t) - \pi/2$.

Therefore, by Theorem 4.3.3, there is at least a nonpositive solution $u$ of (4.4.1) with $\arctan(t) - \pi/2 \le u(t) \le 0, \forall t \in \mathbb{R}$.

Note that the null function is not a solution for the problem and $f$ is discontinuous on $t$ (Fig. 4.4.1).

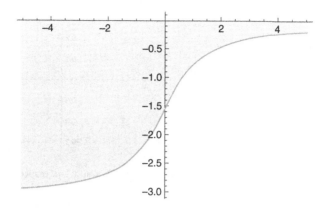

Fig. 4.4.1.   Admissible region for solution $u$.

## 4.5.   Duffing equation

In [8], the authors consider equation

$$-u''(t) + u(t) = a(t)\,|u(t)|^{p-1}\,u(t), \quad t \in \mathbb{R}, \tag{4.5.1}$$

with $p > 1$, which models the forced vibrations of a cantilever beam in the nonuniform field of two permanent magnets.

The structure and behavior of function $a : \mathbb{R} \to \mathbb{R}$ is a key point for the existence of homoclinic solutions. Applying the main result, it can be proved that there exists at least one nontrivial solution in cases not covered, as far as we know, by results in the existent literature.

For example, if $a(t) = -\frac{1}{1+t^2}, p = 3, k = 0.1$, then let us seek a nontrivial and homoclinic solution for

$$\begin{cases} u''(t) - 0.1u(t) = \dfrac{|u(t)|^2\,u(t)}{1+t^2}, & t \in \mathbb{R}, \\[2mm] u(\pm\infty) = u'(\pm\infty) = 0. \end{cases} \tag{4.5.2}$$

The nonlinear function $f : \mathbb{R}^2 \to \mathbb{R}$ defined by

$$f(t, x) = \frac{|x|^2\,x}{1+t^2}$$

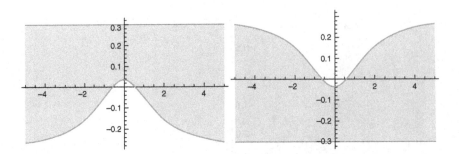

Fig. 4.5.1.   Admissible regions for both solutions $u$ and $-u$, respectively.

is an $L^1$-Carathéodory function with $|x| < \rho$ and $\varphi_\rho(t) = \frac{\rho^3}{1+t^2}$. Functions $\alpha(t) = \frac{1}{3+t^2} - 0.3$ and $\beta(t) \equiv 0.3$ are lower and upper solutions, respectively, of problem (4.5.2).

Therefore, by Theorem 4.3.2, there are at least two homoclinic solutions: $u$ of (4.5.2) and $-u$ of problem

$$\begin{cases} -u''(t) + 0.1u(t) = \dfrac{|u(t)|^2\, u(t)}{1 + t^2}, & t \in \mathbb{R}, \\[2mm] u(\pm\infty) = u'(\pm\infty) = 0 \end{cases} \qquad (4.5.3)$$

with $\frac{1}{3+t^2} - 0.3 \le u(t) \le 0.3$, and $-0.3 \le -u(t) \le -\frac{1}{3+t^2} + 0.3$ for $t \in \mathbb{R}$.

Note that the null function is not a solution, and therefore, $u$ and $-u$ are nontrivial solutions (Fig. 4.5.1).

## 4.6.   Forced cantilever beam equation with damping

The second-order differential equation

$$x''(t) + bx'(t) - x + x^3 = F\cos(\omega t) \qquad (4.6.1)$$

can model the forced vibrations of a cantilever beam in a nonuniform field of two magnets.

As illustrated in Fig. 4.6.1, a slender steel beam is clamped in a rigid framework which supports the magnets. Their attractive forces overcome the elastic ones, which would otherwise keep the beam straight. In the absence of some external force, the beam settles with its tip close to one or the other of the magnets. The variable $x$ represents a measure of the beam's position, say its tip displacement.

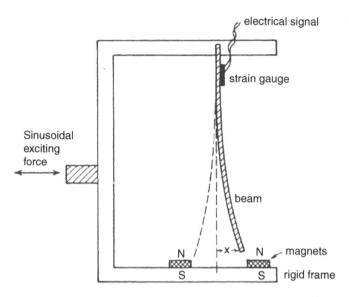

Fig. 4.6.1. Interaction between a cantilever beam, two magnets and an excitation force.

As an example, let us consider the following equation:

$$u''(t) + b(t)u'(t) + c(t)g(t, u) = 0, \qquad (4.6.2)$$

with $b(t) = -\frac{0.01}{1+t^2}, c(t) = 1, g(t, u) = -u - \frac{100u^4}{1+t^2}$.

This class of ODE arises in diffusion phenomena in biomathematics. For more details, see [15, 92].

Note that in this case, the BVP

$$\begin{cases} u''(t) - u(t) = \dfrac{0.01u'(t) + 100u^4(t)}{1+t^2}, & t \in \mathbb{R}, \\ u(\pm\infty) = u'(\pm\infty) = 0 \end{cases} \qquad (4.6.3)$$

is not covered by any kind of existent results to the best of our knowledge.

The nonlinear function $f : \mathbb{R}^3 \to \mathbb{R}$ defined by

$$f(t, x, y) = \frac{0.01y + 100x^4}{1+t^2}$$

is monotone in $y$ for $(t, x) \in \mathbb{R}^2$ fixed, and for $|x|, |y| < \rho$, it is an $L^1$-Carathéodory function with $\varphi_\rho(t) = \frac{0.01\rho + \rho^4}{1+t^2}$.

Functions $\alpha(t) = \frac{1}{1+t^2}$ and $\beta(t) \equiv 0.5$ are, respectively, lower and upper solutions of problem (4.6.3), according to Definition 4.2.6, with $\bar{\alpha}(t) = \frac{1}{1+t^2} - 0.5$.

Therefore, by Theorem 4.3.3, there is at least a homoclinic solution $u$ of (4.6.3) such that

$$\frac{1}{1+t^2} - 0.5 \leq u(t) \leq 0.5, \quad \forall t \in \mathbb{R}.$$

Chapter 5

# Homoclinic Solutions
# to Fourth-Order Problems

## 5.1. Introduction

This chapter provides sufficient conditions for the existence of homoclinic
solutions of fourth-order nonlinear ODEs. Different applications are presented to illustrate new results, such as the nonlinear Bernoulli–Euler–v.
Karman problem, Extended Fisher–Kolmogorov problem and the Swift–
Hohenberg problem. The method will use Green's functions to formulate a
new modified integral equation which is equivalent to the original nonlinear one. Moreover, in an adequate function space, the corresponding nonlinear integral operator is compact, and an existence result by Schauder's
fixed-point theorem can be applied.

We study the existence of homoclinic solutions to the fourth-order nonlinear differential equation

$$u^{(iv)}(t) + ku(t) = f(t, u(t), u'(t), u''(t), u'''(t)), \quad t \in \mathbb{R}, \qquad (5.1.1)$$

with $k > 0$ and $f : \mathbb{R}^5 \to \mathbb{R}$ a continuous function, verifying an adequate
asymptotic condition.

Note that no further condition will be necessary on the nonlinearity
$f(t, x, y, z, w)$ to obtain the existence of homoclinic orbits to 0, that is,
nontrivial solutions of (5.1.1) such that

$$u(\pm\infty) := \lim_{t\to\pm\infty} u(t) = 0, \; u'(\pm\infty) := \lim_{t\to\pm\infty} u'(t) = 0. \qquad (5.1.2)$$

In the last decades, the study of autonomous and nonautonomous
fourth-order differential equations attracted many researchers. To be more

precise, equations of the type

$$u^{(iv)}(t) + ku''(t) + g(u(t)) = 0, \quad t \in \mathbb{R}, \tag{5.1.3}$$

with $k \in \mathbb{R}$, and $g$ a locally Lipschitz function, arise in several theoretical cases and real phenomena such as:

- if $k < 0$, it is known as the *Extended Fisher–Kolmogorov equation* and if $k > 0$, it is referred to as *Swift–Hohenberg equation* (see [126]);
- if $g(u) = u - u^2$, it is applied in the dynamic phase-space analogy of a nonlinearly supported elastic strut (see [83]);
- if $g(u) = u^3 - u$, it models the pattern formation in many physical, chemical or biological systems (see [27]);
- if $g(u) = u^5 - u^3 + u$, it is used to study the localization and spreading of deformation of a strut confined by an elastic foundation (see [11, 125]);
- if $g(u) = (u+1)^+ - 1$, where $(u+1)^+ = \max\{u+1, 0\}$, equation (5.1.3) arises in the search of traveling waves solutions [132] in the study of deflection in railway tracks [1] and undersea pipelines [26].

The existence of homoclinic solutions was proved by using several methods and techniques. Some examples, without pretending to be exhaustive, are shown in [134], where the above nonlinearities by variational arguments and the Palais–Smale condition are considered.

For equation

$$u^{(iv)}(t) + ku''(t) + a(t)u(t) - b(t)u^2(t) - c(t)u^3(t) = 0,$$

the existence of one nontrivial homoclinic solution is proved in [138] with $a(t)$ and $c(t)$ positive bounded and continuous functions, and $b(t)$ a bounded continuous function, applying Mountain Pass Theorem, and the existence of nontrivial homoclinic solutions in the nonperiodic case is proved in [97]. In [89], the authors show the existence of two homoclinic solutions for some nonperiodic fourth-order equations with a perturbation.

This chapter emphasizes on a perturbation with an unknown function where the nonlinearity is given by a generic continuous function with dependence on $u$ and all derivatives till order three.

As far as we know, it is the first time where such perturbation associated to generic nonlinearity, which has to verify only an asymptotic condition, is considered (see assumption (5.3.1)).

The arguments are based in the explicit form of the Green's functions associated to the linear perturbation of (5.1.1) in a compactness criterion and fixed point theory.

## 5.2. Definitions and auxiliary results

Let us define the space

$$X_{H4} = \left\{ x \in C^3(\mathbb{R}) : \lim_{|t| \to +\infty} x(t) = 0, \lim_{|t| \to +\infty} x^{(i)}(t) \in \mathbb{R}, i = 1, 2, 3 \right\}$$

with the norm $\|x\|_{X_{H4}} = \max\{\|x\|_\infty, \|x'\|_\infty, \|x''\|_\infty, \|x'''\|_\infty\}$, where $\|\omega\|_\infty = \sup_{t \in \mathbb{R}} |\omega(t)|$.

In this way, $(X_{H4}, \|\cdot\|_{X_{H4}})$ is a Banach space.

**Remark 5.2.1.** Note that if $u \in X$, then

$$\lim_{|t| \to \infty} u^{(j)}(t) = 0, \quad j = 1, 2, 3.$$

By $u$ solution of problem (9.1.1),(10.3.2), we mean $u \in X$ such that $u$ verifies (9.1.1).

The following result will play an important role in the proof of the main result, giving a solution of some linear fourth-order problem via Green's functions.

**Lemma 5.2.2.** *If $h \in L^1(\mathbb{R})$, then, for some $k > 0$, the problem*

$$\begin{cases} u^{(iv)}(t) + ku(t) = h(t), \ t \in \mathbb{R}, \\ u(\pm\infty) = u'(\pm\infty) = 0 \end{cases} \tag{5.2.1}$$

*has a unique solution in $X_{H4}$. Moreover, this solution can be expressed as*

$$u(t) = \int_{-\infty}^{+\infty} G(t, s)h(s)ds, \tag{5.2.2}$$

*where*

$$G(t, s) = \frac{\sqrt[4]{k}}{2k} e^{-\frac{\sqrt[4]{k}|s-t|}{\sqrt{2}}} \sin\left(\frac{\sqrt[4]{k}|s - t|}{\sqrt{2}} + \frac{\pi}{4}\right). \tag{5.2.3}$$

**Proof.** The homogeneous solution of the linear equation is given by

$$u(t) = e^{At} \left(c_1 \cos(At) + c_2 \sin(At)\right) + e^{-At} \left(c_3 \cos(At) + c_4 \sin(At)\right)$$

with $A = \sqrt[4]{\frac{k}{4}}$ and $c_1, c_2, c_3, c_4 \in \mathbb{R}$.

As the null function is the only solution of the homogeneous problem, Green's functions can be defined and the general solution of (5.2.1) is given by

$$u(t) = \frac{\sqrt[4]{k}}{2k} \int_{-\infty}^{+\infty} e^{-\sqrt[4]{\frac{k}{4}}|s-t|} \sin\left(\sqrt[4]{\frac{k}{4}}|s-t| + \frac{\pi}{4}\right) h(s) ds.$$

For $G(t,s) := \frac{\sqrt[4]{k}}{2k} e^{-A|s-t|} \sin\left(A|s-t| + \frac{\pi}{4}\right)$, one can write

$$u(t) = \int_{-\infty}^{+\infty} G(t,s) h(s) ds.$$

$\square$

The following properties of the Green function can easily be proved.

**Remark 5.2.3.** For $i = 0, 1, 2, 3$, defining

$$G_i^-(t,s) := \frac{\sqrt[4]{k}^{i+1}}{2k} e^{-\frac{\sqrt[4]{k}(s-t)}{\sqrt{2}}} \sin\left(\frac{\sqrt[4]{k}(s-t)}{\sqrt{2}} + \frac{\pi(3i+1)}{4}\right),$$

$$G_i^+(t,s) := \frac{\sqrt[4]{k}^{i+1}}{2k} e^{-\frac{\sqrt[4]{k}(t-s)}{\sqrt{2}}} \sin\left(\frac{\sqrt[4]{k}(t-s)}{\sqrt{2}} + \frac{\pi(3i+1)}{4}\right),$$

then, for $j = 0, 1, 2, 3$,

$$u^{(j)}(t) = \int_{-\infty}^{t} G_j^-(t,s) h(s) ds + (-1)^j \int_{t}^{+\infty} G_j^+(t,s) h(s) ds, \qquad (5.2.4)$$

$$\lim_{|t| \to \infty} \frac{\partial^j G(t,s)}{\partial t^j} = 0, \qquad (5.2.5)$$

$$\left| \frac{\partial^j G(t,s)}{\partial t^j} \right| \leq \frac{(\sqrt[4]{k})^{j+1}}{2k}. \qquad (5.2.6)$$

The following theorem is a key argument to deal with the lack of compactness of the set $X_{H4}$.

**Theorem 5.2.4 ([51]).** *Let $M \subset (C_l, \mathbb{R})$ with*

$$C_l := \left\{ x \in C[0, +\infty) : \text{there exists } \lim_{t \to +\infty} x(t) \right\}.$$

*Then $M$ is compact if the following conditions hold:*

(i) *$M$ is bounded in $C_l$;*
(ii) *functions $f \in M$ are equicontinuous on any compact interval of $[0, +\infty)$;*

(iii) *functions from M are equiconvergent, that is, given $\epsilon > 0$, there exists $T(\epsilon) > 0$ such that $|f(t) - f(+\infty)| < \epsilon$ for all $t > T(\epsilon)$ and $f \in M$.*

The proof of this result can easily be applied to compact intervals of the form $[-T, T]$, for some $T > 0$, as it is suggested in [128], to obtain a similar result to the set $X_{H4}$.

**Theorem 5.2.5.** *A set $M \subset X_{H4}$ is relatively compact if the following conditions hold:*

(i) *$M$ is bounded in $X_{H4}$;*
(ii) *the functions belonging to $M$ are equicontinuous on any compact interval of $\mathbb{R}$;*
(iii) *the functions from $M$ are equiconvergent at $\pm\infty$, that is, given $\epsilon > 0$, there exists $T(\epsilon) > 0$ such that $|f^{(i)}(t) - f^{(i)}(\pm\infty)| < \epsilon$, for all $|t| > T(\epsilon), i = 0, 1, 2, 3$ and $f \in M$.*

## 5.3. Existence results

This section contains an existent result for homoclinic solutions of problem (5.1.1),(5.1.2) without monotone, periodic or extra assumptions on the nonlinear part.

**Theorem 5.3.1.** *Let $f : \mathbb{R}^5 \to \mathbb{R}$ be a continuous function. If for each $r > 0$ with $\max\{\|x\|_\infty, \|y\|_\infty, \|z\|_\infty, \|w\|_\infty\} < r$, there exists a positive function $\phi_r \in L^1(\mathbb{R})$ such that*

$$|f(t, x, y, z, w)| < \phi_r(t), \tag{5.3.1}$$

*then problem (5.1.1),(5.1.2) has a homoclinic solution $u \in X_{H4}$.*

**Proof.** Define

$$F_u(t) := f(t, u(t), u'(t), u''(t), u'''(t))$$

and consider the operator $T : X_{H4} \to X_{H4}$ given by

$$Tu(t) = \int_{-\infty}^{+\infty} G(t, s) F_u(s) ds$$

with $G(t, s)$ defined by (5.2.3).

As $f$ is a continuous function verifying (5.3.1) and $u \in X_{H4}$, it is obvious that $F_u \in L^1(\mathbb{R})$ and, by Lemma 5.2.2, the fixed points of $T$ are solutions of problem (5.1.1),(5.1.2). So, it is enough to prove that $T$ has a fixed point.

Clearly, $Tu \in C^3(\mathbb{R})$ and by (5.2.5) and Lebesgue's Dominated Convergence Theorem,

$$\lim_{|t| \to \infty} (Tu)(t) = \int_{-\infty}^{+\infty} \lim_{|t| \to \infty} G(t,s) F_u(s) ds = 0$$

and, for $i = 1, 2, 3$,

$$\lim_{|t| \to \infty} (Tu)^{(i)}(t) = \int_{-\infty}^{+\infty} \lim_{|t| \to \infty} \frac{\partial^{(i)} G(t,s)}{\partial t^i} F_u(s) ds = 0.$$

Therefore, $Tu \in X_{H4}$, and $T : X_{H4} \to X_{H4}$ is well defined.

Now, for any bounded subset $B \subset X_{H4}$ and any $u \in B$ with $\|u\|_{X_{H4}} \leq r_1$, by (5.2.6) and (5.3.1), one has

$$|Tu(t)| \leq \int_{-\infty}^{+\infty} |G(t,s)||F_u(s)| ds \leq \frac{\sqrt[4]{k}}{2k} \int_{-\infty}^{+\infty} \phi_{r_1} < +\infty, \quad \forall t \in \mathbb{R},$$

and, therefore, $\{Tu(t) : Tu \in B\}$ is relatively compact in $\mathbb{R}$.

For $a > 0$ and $t_1, t_2 \in [-a, a]$, one has, as $t_1 \to t_2$,

$$|Tu(t_1) - Tu(t_2)| = \int_{-\infty}^{+\infty} |G(t_1,s) - G(t_2,s)||F_u(s)| ds \longrightarrow 0,$$

and

$$|(Tu)^{(i)}(t_1) - (Tu)^{(i)}(t_2)|$$
$$= \int_{-\infty}^{+\infty} \left| \frac{\partial^{(i)} G}{\partial t^i}(t_1,s) - \frac{\partial^{(i)} G}{\partial t^i}(t_2,s) \right| |F_u(s)| ds \longrightarrow 0, \quad \text{for } i = 0,1,2,3.$$

So, the set $\{u : [a, -a] \to \mathbb{R}\} \subset B$ is equicontinuous.

By the continuity of $f$ for any $\epsilon > 0$, there exist $t_+ > 0$ and $\delta > 0$ such that when $|u(t) - v(t)| \leq \epsilon$, for $t > t_+$, then

$$|F_u(t_+) - F_v(t_+)| \leq \delta.$$

So, for $i = 1, 2, 3$, and by (5.2.6),

$$|(Tu)^{(i)}(t) - (Tv)^{(i)}(t)| = \int_{-\infty}^{+\infty} \left| \frac{\partial^{(i)} G}{\partial t^i}(t,s) \right| |F_u(s) - F_v(s)| ds \longrightarrow 0,$$

as $t \to +\infty$.

Analogously, when $|u(t) - v(t)| \leq \epsilon$, for $t < -t_+$, then

$$|F_u(-t_+) - F_v(-t_+)| \leq \delta.$$

So, $T$ is equiconvergent at $\pm\infty$, and by Theorem 5.2.5, $TB$ is relatively compact.

Consider now a subset $D \subset X_{H4}$ defined as

$$D := \{u \in X_{H4} : \|u\|_{X_{H4}} < r_2\}$$

with

$$r_2 > \max\left\{r, \int_{-\infty}^{+\infty} M\,\phi_r(s)ds\right\},$$

where $r > 0$ is given by (5.3.1) and

$$M := \max\left\{1, \frac{1}{2\sqrt[4]{k^3}}, \frac{1}{2\sqrt{k}}, \frac{1}{2\sqrt[4]{k}}\right\}$$

with $G_3^-(t,s)$ and $G_3^+(t,s)$ given by Remark 5.2.3.

For $t \in \mathbb{R}$, by (5.2.6) and (5.3.1),

$$\|Tu\| = \sup_{t\in\mathbb{R}}\left|\int_{-\infty}^{+\infty} G(t,s)F_u(s)ds\right|$$

$$\leq \int_{-\infty}^{+\infty} \frac{1}{2\sqrt[4]{k^3}}|f(s,u(s),u'(s),u''(s),u'''(s))|ds$$

$$\leq \int_{-\infty}^{+\infty} \frac{1}{2\sqrt[4]{k^3}}\phi_r(s)ds < r_2,$$

$$\|(Tu)^{(i)}\| = \sup_{t\in\mathbb{R}}\left|\int_{-\infty}^{+\infty} \frac{\partial^{(i)}G}{\partial t^i}(t,s)F_u(s)ds\right|$$

$$\leq \int_{-\infty}^{+\infty} \frac{\left(\sqrt[4]{k}\right)^{i+1}}{2k}\phi_r(s)ds < r_2, \quad \text{for } i = 1, 2,$$

and

$$\|(Tu)'''\| = \sup_{t\in\mathbb{R}}\left|\int_{-\infty}^{t} G_3^-(t,s)F_u(s)ds - \int_{t}^{+\infty} G_3^+(t,s)F_u(s)ds\right|$$

$$\leq \int_{-\infty}^{+\infty} \sup_{t\in\mathbb{R}}\left(|G_3^-(t,s)| + |G_3^-(t,s)|\right)\phi_r(s)ds$$

$$\leq \int_{-\infty}^{+\infty} \phi_r(s)ds < r_2. \tag{5.3.2}$$

Therefore, $TD \subset D$ and, by Theorem 1.2.6, $T$ has at least a fixed point $u \in X_{H4}$. $\qquad\square$

With more information on the asymptotic behavior of the nonlinearity, it is possible to derive more data on solutions of (5.1.1).

**Lemma 5.3.2.** *Let* $k > 0$, $u$ *be a solution of* (5.1.1),(5.1.2) *and* $f$ *be a continuous function verifying*

$$\lim_{\substack{|t| \to +\infty \\ (x,y) \to (0,0)}} f(t, x, y, z, w) = 0. \tag{5.3.3}$$

*Then* $u^{(i)}(\pm\infty) = 0$, $i = 0, 1, 2, 3, 4$.

**Proof.** Let us rewrite equation (5.1.1) as

$$\frac{d}{dt}(e^t(u'''(t) - u''(t) + u'(t) - u(t))) = \delta_1(t)e^t \tag{5.3.4}$$

with $\delta_1(t) = f(t, u(t), u'(t), u''(t), u'''(t)) - (k+1)u(t)$.

By (5.3.3), for any $\epsilon > 0$, there is $\sigma > 0$ such that $|\delta_1(t)| < \epsilon$, for every $t > \sigma$, $|u(t)| < \sigma$, and $|u'(t)| < \sigma$.

Fix $\epsilon > 0$ and integrate (5.3.4) over $]\sigma, t[$, for any $t > \sigma$, to obtain

$$e^t(u'''(t) - u''(t) + u'(t) - u(t)) = C + \int_\sigma^t \delta_1(s)e^s ds,$$

for some $C \in \mathbb{R}$, and, subsequently,

$$|u'''(t) - u''(t) + u'(t) - u(t)| \le |C|e^{-t} + \epsilon e^{-t} \int_\sigma^t e^s ds$$

$$\le |C|e^{-t} + \epsilon(1 - e^{\sigma - t}),$$

for $t > \sigma$.

By letting $t \to +\infty$ and by the arbitrariness of $\epsilon$, the following can be defined:

$$\delta_2(t) := u'''(t) - u''(t) + u'(t) - u(t) \tag{5.3.5}$$

for some continuous function $\delta_2$ vanishing as $t \to +\infty$. Rewriting again equation (5.3.4),

$$\frac{d}{dt}(e^t(u''(t) - 2u'(t) + 3u(t))) := \delta_3(t)e^t \tag{5.3.6}$$

with $\delta_3(t) = \delta_2(t) + 4u(t)$. Arguing as for (5.3.4), it may be defined that

$$\delta_4(t) := u''(t) - 2u'(t) + 3u(t) \tag{5.3.7}$$

for some continuous function $\delta_4(t)$ vanishing as $t \to +\infty$.

Since both $u(t), u'(t) \to 0$, then $u''(t) \to 0$. Similarly, from (5.3.5), it can be demonstrated that $u'''(t) \to 0$, whereas from (5.1.1), $u^{(iv)}(t) \to 0.\square$

## 5.4. Example

Consider the fourth-order BVP

$$\begin{cases} u^{(iv)}(t) + u(t) = \dfrac{u(t)(u''(t) - (u(t))^2) + (u'(t))^2 (u'''(t))^3 + 1}{1 + t^2}, & t \in \mathbb{R}, \\ u(\pm\infty) = u'(\pm\infty) = 0. \end{cases}$$

(5.4.1)

Function $f(t, x, y, z, w) = \frac{x(z - x^2) + y^2 w^3 + 1}{1 + t^2}$ is continuous and verifies (5.3.1) for $\max\{\|x\|_\infty, \|y\|_\infty, \|z\|_\infty, \|w\|_\infty\} < r_1, (r_1 > 0)$ with

$$\phi_{r_1}(t) := \frac{r_1^2 + r_1^3 + r_1^5 + 1}{1 + t^2}.$$

Therefore, by Theorem 5.3.1 there exists a nonnegative homoclinic solution of problem (5.4.1) with the phase portrait and its graphic given by Figs. 5.4.1 and 5.4.2.

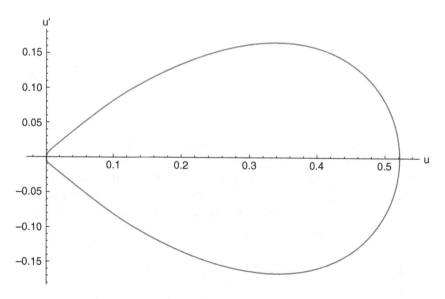

Fig. 5.4.1.　Phase portrait of the homoclinic solution $u$ of (5.4.1).

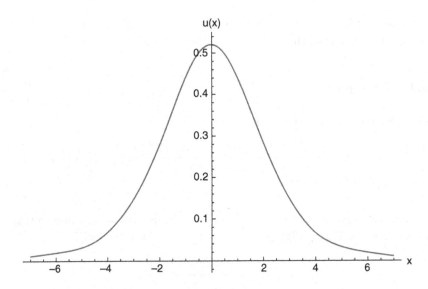

Fig. 5.4.2.   Graph of the homoclinic solution $u$ of (5.4.1).

## 5.5.   Bernoulli–Euler–v. Karman problem

In [87], the nonlinear Bernoulli–Euler–v. Karman BVP is considered:

$$\begin{cases} EIu^{(iv)}(t) + ku(t) = \frac{3}{2}EA(u'(t))^2 u''(t) + \omega(t), \ t \in \mathbb{R}, \\ u(\pm\infty) = u'(\pm\infty) = 0, \end{cases} \tag{5.5.1}$$

which is related to the analysis of moderately large deflections of infinite nonlinear beams resting on elastic foundations under localized external loads. More precisely, $E$ is the Young's modulus, $I$ the mass moment of inertia, $ku(t)$ the spring force upward, in which $k$ is a spring constant (for simplicity, the weight of the beam is neglected), $A$ the cross-sectional area of the beam and $\omega(t)$ the applied loading downward (see Fig. 5.5.1). An example of this family of problems is given by

$$\begin{cases} u^{(iv)}(t) + 3u(t) = \dfrac{3.4 + u^3(t) - u''(t)\left(u'(t)\right)^2}{1 + t^4}, \quad t \in \mathbb{R}, \\ u(\pm\infty) = u'(\pm\infty) = 0. \end{cases} \tag{5.5.2}$$

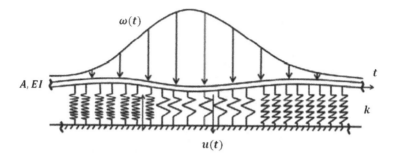

Fig. 5.5.1. Infinite nonlinear beam resting on nonuniform elastic foundations.

Here, the loading force $\omega(t) = \frac{3.4}{1+t^4}$ and the nonlinear function $g : \mathbb{R}^4 \to \mathbb{R}$ is defined by

$$g(t, x, y, z) := \frac{x^3 - zy^2}{1 + t^4}.$$

The function $f(t, x, y, z) := g(t, x, y, z) + \omega(t)$ is continuous and verifies (5.3.1) for $\max\{\|x\|_\infty, \|y\|_\infty, \|z\|_\infty\} < r_2, (r_2 > 0)$ with

$$\phi_{r_2}(t) := \frac{3.4 + 2r_2^3}{1 + t^4}.$$

By Theorem 5.3.1, there is a nontrivial homoclinic solution $u^*$. Moreover, as $f$ verifies (5.3.3), by Corollary 5.3.2, this homoclinic solution $u^*$ of (5.5.2) verifies $(u^*)^{(i)}(\pm\infty) = 0$ for $i = 0, 1, 2, 3, 4$.

## 5.6. Extended Fisher–Kolmogorov and Swift–Hohenberg problems

In [89], the authors consider a fourth-order differential equation which can be written as

$$u^{(iv)}(t) + u(t) = 2u(t) - au''(t) - u^3(t), \quad t \in \mathbb{R}. \tag{5.6.1}$$

In the literature, when $a < 0$, this equation corresponds to the well-known Extended Fisher–Kolmogorov (EFK) equation, proposed in [52], to study phase transitions. If $a > 0$, equation (5.6.1) is related to Swift–Hohenberg (SH) equation, which is a general model for pattern-forming

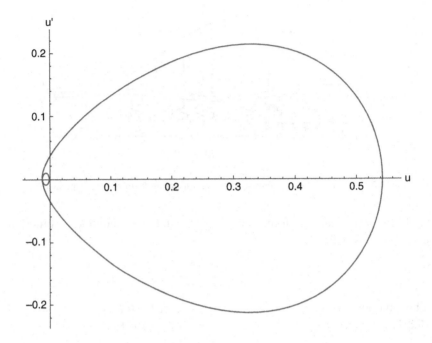

Fig. 5.6.1.  Phase portrait of the homoclinic solution of (5.6.2), (5.1.2).

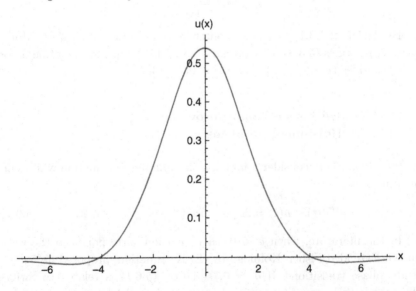

Fig. 5.6.2.  Graph of the homoclinic solution of (5.6.2), (5.1.2).

process, to describe random thermal fluctuations in the Boussinesq equation (see [137]) and in the propagation of lasers (see [96]).

In this sense, equation

$$u^{(iv)}(t) + u(t) = \frac{(1 + u(t))\left(1 + u''(t) - u^2(t)\right)}{1 + t^4}, \quad t \in \mathbb{R} \qquad (5.6.2)$$

can be seen as a generalized (EFK), or (SH), where the coefficient of $u''(t)$ depends on the unknown function and it does not have a definite signal. In both cases of the coefficient sign, the nonlinear function $f : \mathbb{R}^3 \to \mathbb{R}$ defined by

$$f(t, x, z) := \frac{(1 + x)(1 + z - x^2)}{1 + t^4}$$

is continuous and for $\max\{\|x\|_\infty, \|z\|_\infty\} < r_3$, $(r_3 > 0)$, $f$ verifies (5.3.1) with

$$\phi_{r_3}(t) := \frac{(1 + r_3)(1 + r_3 + r_3^2)}{1 + t^4}.$$

Therefore, by Theorem 5.3.1, there is a homoclinic solution $u^*$ of problem (5.6.2),(5.1.2). As illustrated in Figs. 5.6.1 and 5.6.2, this homoclinic solution is a sign-changing function.

# Chapter 6

# Lidstone Boundary Value Problems

## 6.1. Introduction

George James Lidstone (1870–1952) was an English mathematician who worked, among other things, on the study of polynomial interpolation. In 1929, he introduced a generalization of Taylor's series, where the innovation part was an approximation of a given function in the neighborhood of two points instead of one.

Essentially, this interpolating polynomial is a solution of a BVP given by an elementary even-order differential equation and boundary conditions defined on a bounded interval

$$\begin{cases} u^{(2m)}(t) = 0, \quad t \in [a, b], \\ u^{(j)}(a) = A_j, u^{(j)}(b) = B_j, \quad j = 0, 1, \ldots, m - 1. \end{cases}$$

In the field of approximation theory, the Lidstone interpolating polynomial of degree $(2m - 1)$ matches $u(t)$ and its $(m - 1)$ even derivatives at both ends of the compact interval.

The homogeneous differential equation can be generalized and, coupled with boundary conditions, generates the next BVP

$$\begin{cases} u^{(2m)}(t) = f(t, u(t), u'(t), \ldots, u^{(2m-1)}(t)), t \in [0, 1], \\ u^{(j)}(0) = A_j, u^{(j)}(1) = B_j, j = 0, 1, \ldots, m - 1. \end{cases}$$

This kind of BVP is known as *Lidstone boundary value problems*.

The particular case $m = 2$ frequently occurs in engineering and other branches of physical sciences. For instance, the deflection of a uniformly loaded rectangular plate, supported over the entire surface by an elastic foundation and rigidly supported along the edges, leads to this type of problem, or modeling the deformations of an elastic beam where the type of boundary conditions considered depends on how the beam is supported at the two endpoints (see [77] and the references therein).

In this specific case, Lidstone boundary conditions,

$$u(a) = u''(a) = u(b) = u''(b) = 0,$$

mean that both endpoints of the beam are simply supported.

Recently, it was introduced the so-called *complementary Lidstone boundary value problems* (see [6, 7, 143]) with differential equations of odd order together with odd boundary derivatives conditions only, of the following type, were introduced:

$$\begin{cases} u^{(2m-1)}(t) = f(t, u(t), u'(t), \ldots, u^{(2m-2)}(t)), t \in [a, b], \\ u(a) = A_0, \ u^{(2j-1)}(a) = A_j, u^{(2j-1)}(b) = B_j, j = 1, \ldots, m. \end{cases}$$

These types of problems with full nonlinearities, that is, with dependence on even and odd derivatives, are very scarce (see [62, 64, 119]). However, as far as we know, Lidstone or complementary Lidstone problems were never applied to the whole real line.

This chapter is concerned with the study of a fully nonlinear differential equation on the real line

$$u^{(iv)}(t) + k^4 u(t) = f(t, u(t), u'(t), u''(t), u'''(t)), \quad t \in \mathbb{R}, \qquad (6.1.1)$$

where $k \in \mathbb{R}, f : \mathbb{R}^5 \to \mathbb{R}$ is a continuous function and two Lidstone-type boundary conditions: the classical ones, with even derivatives,

$$u(\pm\infty) = u''(\pm\infty) = 0, \qquad (6.1.2)$$

with $u^{(i)}(\pm\infty) := \lim_{t \to \pm\infty} u^{(i)}(t), i = 0, 2$ and the so-called complementary Lidstone boundary conditions

$$u(\pm\infty) = u'(\pm\infty) = 0. \qquad (6.1.3)$$

Note that solutions of problem (6.1.1),(6.1.3) are homoclinic solutions and in this way, the results of this chapter complement and generalize the ones achieved in Chapter 5.

The main arguments are based on the explicit form of Green's functions associated to problem (6.1.1),(6.1.2) in a compactness criterion and fixed point theory.

The problem (6.1.1),(6.1.2) can model several real phenomena in beam theory (see [1, 16]), suspension bridges (see [13, 22]) and elasticity theory, among others. Equation (6.1.1) is often referred to as a beam equation because it describes the deflection of an elastic beam under a certain force. The boundary conditions (6.1.2) mean that the beam is simply supported at infinity.

## 6.2. Auxiliary definitions and Green's functions

The space of admissible functions to be used forward will be

$$X_L = \left\{ x \in C^3(\mathbb{R}) : \lim_{|t| \to +\infty} x(t) = 0 \right\},$$

equipped with the norm $\|x\|_{X_L} = \max\{\|x\|_\infty, \|x'\|_\infty, \|x''\|_\infty, \|x'''\|_\infty\}$, where $\|\omega\|_\infty = \sup_{t \in \mathbb{R}} |\omega(t)|$.

In this way, $(X_L, \|\cdot\|_{X_L})$ is a Banach space.

The following result will play an important role in the proof of the main result, giving an explicit solution of some linear fourth-order problem via Green's functions.

**Lemma 6.2.1.** *If* $h \in L^1(\mathbb{R})$, *then for* $k \in \mathbb{R}$, *the linear problem*

$$\begin{cases} u^{iv}(t) + k^4 u(t) = h(t), t \in \mathbb{R}, \\ u(\pm\infty) = u''(\pm\infty) = 0 \end{cases} \tag{6.2.1}$$

*has a unique solution in* $X_L$ *which can be expressed as*

$$u(t) = \int_{-\infty}^{+\infty} G(t, s) h(s) ds,$$

*where*

$$G(t, s) = \frac{e^{-k_*|s-t|}}{\sqrt{2}^5 k_*^3} \sin\left(k_*|s - t| + \frac{\pi}{4}\right) \tag{6.2.2}$$

*with* $k_* = \frac{k\sqrt{2}}{2}$.

**Proof.** The homogeneous solution of the linear equation is given by

$$u(t) = e^{k_* t} \left( c_1 \cos(k_* t) + c_2 \sin(k_* t) \right) + e^{-k_* t} \left( c_3 \cos(k_* t) + c_4 \sin(k_* t) \right)$$

with $c_1, c_2, c_3, c_4 \in \mathbb{R}$ and the general solution of the homogeneous problem associated to (6.2.1) is given by

$$u(t) = \frac{1}{2k^3} \int_{-\infty}^{+\infty} e^{-k_* |s-t|} \sin\left( k_* |s - t| + \frac{\pi}{4} \right) h(s) ds.$$

For $G(t,s) := G(t,s) = \frac{e^{-k_* |s-t|}}{\sqrt{2}^5 k_*^3} \sin(k_* |s - t| + \frac{\pi}{4})$, one can write

$$u(t) = \int_{-\infty}^{+\infty} G(t,s) h(s) ds.$$

$\square$

Some properties of these Green's functions are in the following remark.

**Remark 6.2.2.** For $i = 0, 1, 2, 3$, defining

$$G_i^-(t,s) := \frac{e^{k_*(s-t)}}{\sqrt{2}^{5-i} k_*^{3-i}} \sin\left( k_*(t - s) + \frac{\pi(3i + 1)}{4} \right),$$

$$G_i^+(t,s) := \frac{e^{k_*(t-s)}}{\sqrt{2}^{5-i} k_*^{3-i}} \sin\left( k_*(s - t) + \frac{\pi(3i + 1)}{4} \right),$$

one has

$$u^{(i)}(t) = \int_{-\infty}^{t} G_i^-(t,s) h(s) ds + (-1)^i \int_{t}^{+\infty} G_i^+(t,s) h(s) ds. \qquad (6.2.3)$$

The following properties of the Green function can easily be proved:

$$\lim_{|t| \to +\infty} G(t,s) = \lim_{t \to +\infty} G_i^-(t,s) = \lim_{t \to -\infty} G_i^+(t,s) = 0, \qquad (6.2.4)$$

$$|G_i(t,s)| \le \frac{1}{\sqrt{2}^{5-i} k_*^{3-i}}, \qquad i = 0, 1, 2, 3. \qquad (6.2.5)$$

The following theorem is a key argument to deal with the lack of compactness.

**Theorem 6.2.3.** *For a set $D \subset X_L$ to be relatively compact, it is necessary and sufficient that*

(i) *$\{x(t) : x \in D\}$ is relatively compact in $\mathbb{R}$ for any $t \in \mathbb{R}$;*
(ii) *for each $a > 0$, the family $D_a := \{x : [-a, a] \to \mathbb{R}\} \subset D$ is equicontinuous;*

(iii) *D is stable at ±∞, i.e., for arbitrary functions x and y in D, and any*
$\epsilon > 0$, *there exist* $T > 0$ *and* $\delta > 0$, *such that if* $|x^{(i)}(T) - y^{(i)}(T)| \leq \delta$,
*then* $|x^{(i)}(t) - y^{(i)}(t)| \leq \epsilon$ *for* $t > T$, *and if* $|x^{(i)}(-T) - y^{(i)}(-T)| \leq \delta$,
*then* $|x^{(i)}(t) - y^{(i)}(t)| \leq \epsilon$ *for* $t < -T$ *for each* $i = 0, 1, 2, 3$.

**Proof.** The proof is a direct application of [128, Theorem 1]. □

## 6.3. Existence result

The main result of this chapter is given by the following theorem.

**Theorem 6.3.1.** *Let* $f : \mathbb{R}^5 \to \mathbb{R}$ *be a continuous function. If for each*
$r > 0$ *with* $\max\{\|x\|_\infty, \|y\|_\infty, \|z\|_\infty, \|w\|_\infty\} < r$, *there exists a positive*
*function* $\phi_r : \mathbb{R} \to [0, +\infty)$ *such that*

$$|f(t, x, y, z, w)| < \phi_r(t) \quad and \quad \int_{-\infty}^{+\infty} \phi_r(t)dt < +\infty, \qquad (6.3.1)$$

*then problem* (6.1.1),(6.1.2) *has a solution* $u \in X_L$, *which is also a homoclinic solution.*

**Proof.** Define

$$F_u(t) := f(t, u(t), u'(t), u''(t), u'''(t)),$$

and consider the operator $T : X_L \to X_L$ given by

$$Tu(t) = \int_{-\infty}^{+\infty} G(t, s)F_u(s)ds$$

with $G(t, s)$ defined by (6.2.2).

As $f$ is a continuous function, $u \in X_L$, and verifies (6.3.1), it is obvious that $F_u \in L^1(\mathbb{R})$, and, by Lemma 6.2.1, fixed points of $T$ are solutions of problem (6.1.1),(6.1.2). So, it is enough to prove that $T$ has a fixed point.

Clearly, $Tu \in C^3(\mathbb{R})$ and, by Lebesgue's Dominated Convergence Theorem and (6.2.4),

$$\lim_{|t| \to +\infty} (Tu)(t) = \int_{-\infty}^{+\infty} \lim_{|t| \to +\infty} G(t, s)F_u(s)ds = 0,$$

$$\lim_{|t| \to +\infty} (Tu)''(t) = \int_{-\infty}^{+\infty} \lim_{|t| \to +\infty} G_2(t, s)F_u(s)ds = 0,$$

and

$$\lim_{|t| \to +\infty} (Tu)'(t) = \int_{-\infty}^{t} \lim_{t \to +\infty} G_1^- F_u(s) ds - \int_{t}^{+\infty} \lim_{t \to -\infty} G_1^+ F_u(s) ds = 0,$$

$$\lim_{|t| \to +\infty} (Tu)'''(t) = \int_{-\infty}^{t} \lim_{t \to +\infty} G_3^- F_u(s) ds - \int_{t}^{+\infty} \lim_{t \to -\infty} G_3^+ F_u(s) ds = 0.$$

Therefore, $Tu \in X_L$, and $T : X_L \to X_L$ is well defined.

Let $B \subset X_L$ be a bounded subset, that is, there is $r_1 > 0$ such that, for any $u \in B$, one has $\|u\|_{X_L} < r_1$. By (6.2.5) and (6.3.1), for $i = 0, 1, 2, 3$,

$$|(Tu(t))^{(i)}| \le \int_{-\infty}^{+\infty} |G_i(t, s)| |F_u(s)| ds$$

$$\le \frac{1}{\sqrt{2}^{5-i} k_*^{3-i}} \int_{-\infty}^{+\infty} \phi_{r_1}(s) ds < +\infty, \quad \forall t \in \mathbb{R},$$

and therefore, $\{Tu(t) : Tu \in B\}$ is relatively compact in $\mathbb{R}$.

For some $a > 0$ and $t_1, t_2 \in [-a, a]$, as $t_1 \to t_2$,

$$|Tu(t_1) - Tu(t_2)| = \int_{-\infty}^{+\infty} |G(t_1, s) - G(t_2, s)| |F_u(s)| ds \longrightarrow 0,$$

$$|(Tu)''(t_1) - (Tu)''(t_2)| = \int_{-\infty}^{+\infty} |G_2(t_1, s) - G_2(t_2, s)| |F_u(s)| ds \longrightarrow 0,$$

and for $i = 1, 3$,

$$\int_{-\infty}^{t} |G_i^-(t_1, s) - G_i^-(t_2, s)| |F_u(s)| ds$$

$$+ \int_{t}^{+\infty} |G_i^+(t_1, s) - G_i^+(t_2, s)| |F_u(s)| ds \longrightarrow 0.$$

So, the set $\{u : [-a, a] \to \mathbb{R}\} \subset B$ is equicontinuous.

As the stability at $\pm\infty$, by the continuity of $f$, for any $\epsilon > 0$, there exist $t_+ > 0$ and $\delta > 0$ such that when $|u(t) - v(t)| \le \epsilon$, for $t > t_+$, then

$$|F_u(t_+) - F_v(t_+)| \le \delta.$$

So, for $i = 0, 1, 2, 3,$

$$|(Tu)^{(i)}(t) - (Tv)^{(i)}(t)| \le \int_{-\infty}^{t} |G_i^-(t,s)||F_u(s) - F_v(s)|ds$$

$$+ \int_{t}^{+\infty} |G_i^+(t,s)||F_u(s) - F_v(s)|ds \longrightarrow 0,$$

as $t \to +\infty$.

Analogously, when $|u(t) - v(t)| \le \epsilon$, for $t < -t_+$, then

$$|F_u(-t_+) - F_v(-t_+)| \le \delta.$$

So, $T$ is stable at $\pm\infty$, and by Theorem 6.2.3, $TB$ is relatively compact.

Consider now a subset $D \subset X_L$ defined as

$$D := \{u \in X_L : \|u\|_{X_L} \le r_2\}$$

with

$$r_2 > \max\left\{r, \int_{-\infty}^{+\infty} M \,\phi_r(s)ds\right\},$$

where $r > 0$ is given by (6.3.1) and

$$M := \max\left\{1, \frac{1}{\sqrt{2}^5 k_*^3}, \frac{1}{2k_*^2}, \frac{1}{\sqrt{2}^3 k_*}\right\}.$$

For $t \in \mathbb{R}$, by (6.2.5) and (6.3.1),

$$\|Tu\|_\infty = \sup_{t\in\mathbb{R}} \left|\int_{-\infty}^{+\infty} G(t,s)F_u(s)ds\right|$$

$$\le \int_{-\infty}^{+\infty} \frac{1}{\sqrt{2}^5 k_*^3} |f(s, u(s), u'(s), u''(s), u'''(s))| \, ds$$

$$\le \int_{-\infty}^{+\infty} \frac{1}{\sqrt{2}^5 k_*^3} \phi_r(s)ds < r_2,$$

$$\|(Tu)''\|_\infty = \sup_{t\in\mathbb{R}} \left|\int_{-\infty}^{+\infty} G_2(t,s)F_u(s)ds\right| \le \int_{-\infty}^{+\infty} \frac{1}{\sqrt{2}^3 k_*} \phi_r(s)ds < r_2,$$

and

$$\|(Tu)'\|_\infty = \sup_{t\in\mathbb{R}} \left| \int_{-\infty}^{t} G_1^-(t,s)F_u(s)ds - \int_{t}^{+\infty} G_1^+(t,s)F_u(s)ds \right|$$

$$\leq \int_{-\infty}^{+\infty} \sup_{t\in\mathbb{R}} \left( \left|G_1^-(t,s)\right| + \left|G_1^-(t,s)\right| \right) \phi_r(s)ds$$

$$\leq \frac{1}{2k_*^2} \int_{-\infty}^{+\infty} \phi_r(s)ds < r_2,$$

$$\|(Tu)'''\|_\infty = \sup_{t\in\mathbb{R}} \left| \int_{-\infty}^{t} G_3^-(t,s)F_u(s)ds - \int_{t}^{+\infty} G_3^+(t,s)F_u(s)ds \right|$$

$$\leq \int_{-\infty}^{+\infty} \sup_{t\in\mathbb{R}} \left( \left|G_3^-(t,s)\right| + \left|G_3^-(t,s)\right| \right) \phi_r(s)ds$$

$$\leq \int_{-\infty}^{+\infty} \phi_r(s)ds < r_2.$$

Therefore, $TD \subset D$ and, by Theorem 1.2.6, $T$ has at least a fixed point $u \in X_L$.

This fixed point is a solution of (6.1.1),(6.1.2) and, moreover, a homoclinic solution of (6.1.1),(6.1.2), by Lemma 4.2.1.  □

**Remark 6.3.2.** By Lemma 4.2.1, the solution of problem (6.1.1),(6.1.2) given by the previous theorem, is also a solution of the complementary Lidstone problem (6.1.1),(6.1.3).

## 6.4.    An infinite beam resting on granular foundations

Soil improvement via stone columns (filling a cylindrical cavity with granular material) is achieved by accelerating the consolidation of the soft soil due to the shortened drainage path with an increase in the load-carrying capacity and/or a decrease in the settlement due to the inclusion of stronger granular material.

Apart from improving the ground below the foundations of residential as well as industrial buildings, stone columns are also installed in soft soils or loose sand for railways and roadways due to the stringent settlement restrictions.

Many studies are available on the analysis of rails, treated as infinite beams on elastic foundations, subjected to concentrated moving loads as

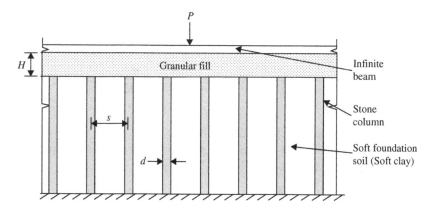

Fig. 6.4.1. Railway beam resting on reinforced granular fill-poor soil system.

well as dynamic loads, using different techniques. For details, see [107, 109, 122] and the references therein.

A longitudinal section of a rail idealized as an infinite beam resting on a ballast layer of a granular fill-stone column-reinforced soft soil system is sketched in Fig. 6.4.1.

The beam is founded on a granular fill layer of thickness $H$ overlying saturated soft soil. The shear modulus of the granular fill layer is $G$. The diameter and the spacing of the stone columns are $d$ and $s$, respectively.

In [107], the differential equation of an infinite beam with a uniform cross-section and a moving load can be written as

$$EI\frac{d^4w}{d\xi^4} + \rho v^2\frac{d^2w}{d\xi^2} - c\frac{dw}{d\xi} + q = P(\xi),$$

where $EI$ is the flexural rigidity of the infinite beam, $\xi$ is the distance from the point of action of load at time $t$ has been considered as $\xi = x - vt$, where $v$ is the constant velocity at which the load is moving on the infinite beam, $w(\xi)$ is the transverse displacement of the beam at $\xi$, $\rho$ is the mass per unit length of the beam, $c$ is the coefficient of viscous damping per unit length of the beam, $P(\xi)$ is the applied load intensity and $q$ is the reaction of the granular fill on the beam, a function that involves the shear modulus $G$ and the thickness of the granular fill layer $H$.

Suppose that, (see [107]),

$$q := \left(1 - \frac{1}{1+\xi^2}\right) a\, w - GH\, \frac{d^2w}{d\xi^2}$$

for some positive parameters $a, b$ and $d$. Then an example of this type of problems is given by the Lidstone boundary value problem in the whole real line, composed by the differential equation

$$\frac{d^4 w}{d\xi^4} + \frac{a}{EI} w = \frac{1}{1+\xi^2} \frac{1}{EI} \left[ (GH - \rho v^2) \frac{d^2 w}{d\xi^2} + cv \frac{dw}{d\xi} + aw + P(\xi) \right],$$

(6.4.1)

together with the boundary conditions (6.1.2).

This problem (6.4.1),(6.1.2) is a particular case of the initial problem (6.1.1),(6.1.2) with $k^4 = \frac{a}{EI}$ and

$$f(\xi, x_1, x_2, x_3, x_4) := \frac{1}{1+\xi^2} \frac{1}{EI} \left[ (GH - \rho v^2) x_3 + cvx_2 + a\, x_1 + P(\xi) \right]$$

is a continuous function.

If the applied load $P(\xi)$ is bounded, that is, there is $K > 0$ such that $\|P\| \leq K$, and not identically to 0, then $f$ verifies (6.3.1) with

$$\phi_r(\xi) := \frac{1}{1+\xi^2} \frac{1}{EI} \left[ |GH - \rho v^2| r + (cv + a) r + K \right].$$

By Theorem 6.3.1, there is a nontrivial solution $w$ of problem (6.4.1),(6.1.2), which is, by Lemma 4.2.1, a homoclinic solution.

Part III

# Heteroclinic Solutions and Hammerstein Equations

# Introduction

This part combines several fields of differential and integral equations, such as heteroclinic connections between two equilibrium points, semi-linear problems, or problems with $\phi$-Laplacian equations, and integral equations of Hammerstein type, all of them defined on the whole real line.

The interest in heteroclinic connections arises in part from the role they play in some models for phase transitions, in particular time-dependent and stationary solutions, that is why heteroclinic solutions are often referred as transitional solutions. The study of sufficient conditions to guarantee the existence of heteroclinic solutions for some boundary value problems has increased in recent years, mainly due to the applications to non-Newtonian fluid theory, diffusion of flows in porous media, nonlinear elasticity and its relations to processes in which the variable transits from an unstable equilibrium to a stable one.

Differential equations including nonlinear differential operators have been widely studied. Perhaps, the most investigated operator is the classical $p$-Laplacian, $\phi_p(y) := y|y|^{p-2}$ with $p > 1$, which, in the recent years, has been generalized to other types of differential operators that preserve the monotonicity of the $p$-Laplacian. These more general operators are usually referred to as $\phi$-Laplacian or semi-linear operators. Therefore, the related nonlinear differential equation has, for a second-order fully differential equation, the form

$$(\phi(u'(t)))' = f(t, u(t), u'(t)),$$

where $\phi : \mathbb{R} \to \mathbb{R}$ is an increasing homeomorphism such that $\phi(0) = 0$. More recently, the case has been considered in which the increasing homeomorphism $\phi$ is defined on the whole real line but is not surjective (see, e.g., [23]), and the case in which $\phi$ is defined only on a bounded domain (see, e.g., [24]). In this case, such an operator is also called singular $\phi$-Laplacian.

As it is known, the main difficulty to pass from $p = 2$ to $p \neq 2$ is the fact that in the first case, when $p = 2$, the differential equation can be written as an equivalent integral equation applying the Green's function technique. However, for $p \neq 2$, it is impossible to find such Green's function in the equivalent integral operator since the differential operator $(\phi(u'))'$ is nonlinear.

The first three chapters of this part present sufficient conditions for three different semi-linear problems, involving general $\phi$-Laplacian equations defined on the whole real line, including for some of them, the singular $\phi$-Laplacian case. Let us point out that, in each case, the existence of heteroclinic solutions is obtained without asymptotic, growth or other extra assumptions on the nonlinearities $\phi$ and $f$. Roughly speaking, our method applies conditions on the inverse operator $\phi^{-1}$, rather than on $\phi$ and $f$, as it is usual in the literature. Moreover, this technique remains useful, even in the case where $\phi(y) = y$.

As it was mentioned in the case of the $p$-Laplacian, a key method to deal with BVP is to write an equivalent integral equation. In this way, we can see the integral equations as generalizations of BVPs. In fact, the nonlinear Hammerstein integral equations have been one of the most important fields of application of the methods and techniques of nonlinear functional analysis and they have been extensively studied since Hammerstein published the seminal paper [80].

Chapter 10 contains a Hammerstein integral equation defined on the real line, where the discontinuous nonlinearity can depend on the derivatives too, without assuming monotone or asymptotic conditions. We point out that the kernel functions, $k(t, s)$, and their partial derivatives in order to the first variable, may be discontinuous and may change signal. Our method presents two features, among others:

- the value of the limit of $k(t, s)$, when $|t| \to \infty$, can be seen as a criterion to classify the existent solutions as homoclinic or heteroclinic solutions;
- it can be applied to boundary value problems with differential equations of any order $n > m$, $m$ being the higher order of derivatives on the nonlinearity.

The last section contains an application to a fourth-order nonlinear boundary value problem, which models moderately large deflections of infinite nonlinear beams resting on elastic foundations under localized external loads.

# Chapter 7

# Heteroclinic Solutions
# for Semi-linear Problems (i)

## 7.1. Introduction

In the recent years, a wide literature has been produced to study boundary value problems (BVPs) composed by differential equations of the form

$$(\phi(u'(t)))' = f(t, u(t), u'(t))$$

with $\phi$ an increasing homeomorphism, and different types of boundary conditions. A classical operator of this family is the $p$-Laplacian $\phi_p(y) = |y|^{p-2}y$, $(p > 1)$, which arises in many models, such as non-Newtonian fluids theory, diffusion of flows in porous media and nonlinear elasticity, among others (see, for example, [14, 23, 42, 59, 108, 123, 127]).

More recently, BVPs on the half or the whole line have been considered with surjective or nonsurjective (singular) homeomorphisms, and sufficient conditions for the existence of homoclinic or heteroclinic solutions were obtained (see, for instance, [9, 43, 85, 101, 111, 135, 138, 140] and the references there in) or for the solvability of problems with integral boundary conditions (see [102, 103]).

In [37], the problem is studied

$$(\phi(u'(t)))' = f(t, u(t), u'(t)), \quad \text{on } \mathbb{R},$$

$$u(-\infty) = -1, \quad u(+\infty) = 1,$$

with the following assumptions on the nonlinearity $f$:

($f$0) $f : \mathbb{R}^3 \to \mathbb{R}$ is continuous and satisfies the symmetry condition

$$f(t, x, y) = -f(-t, -x, y) \quad \text{for all } (t, x, y) \in \mathbb{R}^3;$$

(*f*1) $f(t, 1, y) = 0 = f(t, -1, y)$ for all $t, y \in \mathbb{R}$;

(*f*2) $f(t, x, y) < 0$ for all $t > 0, -1 < x < 1$ and $y \in \mathbb{R}$. Moreover, for every compact set of the form $K = [-r, r] \times [-\varepsilon, \varepsilon]$, where $0 < r < 1$ and $0 < \varepsilon < 1$, there exist $t_K \geq 0$ and a continuous function $h_K : [t_K, \infty) \to \mathbb{R}$ such that

$$f(t, x, y) \leq h_K(t) \quad \text{for all } t \geq t_K \quad \text{and} \quad (x, y) \in K,$$

and

$$\int_{-\infty}^{+\infty} h_K(s)\, ds = -\infty.$$

In [103], the problem is considered

$$(\phi(\rho(t)x'(t)))' + f(t, x(t), x'(t)) = 0, \text{ on } \mathbb{R},$$

$$\lim_{t \to -\infty} x(t) = \int_{-\infty}^{+\infty} g\left(s, x(s), x'(s)\right) ds,$$

$$\lim_{t \to +\infty} x(t) = \int_{-\infty}^{+\infty} h(s, x(s), x'(s)) ds,$$

where

- $\rho \in C(\mathbb{R}, [0, \infty))$ with $\rho(t) \geq 0$ for all $t \in \mathbb{R}$ and satisfying $\int_{-\infty}^{+\infty} \frac{ds}{\rho(s)} < +\infty$;
- $\phi : \mathbb{R} \to \mathbb{R}$ is a strictly increasing sup-multiplicative-like function;
- $f, g, h$ defined on $\mathbb{R}^3$, are Carathéodory functions, verifying some growth conditions, at most linear on the space variables.

In this chapter, we consider the second-order discontinuous equation in the real line,

$$(\phi(a(t)u'(t)))' = f(t, u(t), u'(t)), \quad \text{a.e. } t \in \mathbb{R}, \qquad (7.1.1)$$

with $\phi$ an increasing homeomorphism such that $\phi(0) = 0$ and $\phi(\mathbb{R}) = \mathbb{R}$, $a \in C(\mathbb{R})$ with $a(t) > 0$, for $t \in \mathbb{R}$, and $f : \mathbb{R}^3 \to \mathbb{R}$ an $L^1$-Carathéodory function.

We look for heteroclinic orbits, that is, nontrivial solutions of (7.1.1) such that

$$u(-\infty) := \lim_{t \to -\infty} u(t) = A, u(+\infty) := \lim_{t \to +\infty} u(t) = B, \qquad (7.1.2)$$

where $A, B \in \mathbb{R}$ such that $A < B$.

Remark that the existence of heteroclinic solutions for (7.1.1) is obtained without asymptotic, growth or extra assumptions on the nonlinearities $\phi$ and $f$, applying similar arguments as in [71, 72]. Moreover, this result still holds when $\phi(y) = y$, that is, for equation

$$(a(t)u'(t))' = f(t, u(t), u'(t)), \quad \text{a.e.} \, t \in \mathbb{R}.$$

## 7.2. Definitions and preliminary results

In this section, we present the functional framework for our problem and some auxiliary results.

Consider the space

$$X = \left\{ u \in C^1(\mathbb{R}) : \lim_{|t| \to +\infty} u^{(i)}(t) \in \mathbb{R}, \quad i = 0, 1 \right\}$$

with the norm $\|x\|_X = \max\{\|x\|_\infty, \|x'\|_\infty\}$, where $\|y\|_\infty := \sup_{t \in \mathbb{R}} |y(t)|$.

It is clear that $(X, \| \cdot \|_X)$ is a Banach space.

By a solution of problem (7.1.1),(7.1.2), we mean a function $u \in X$ such that $\phi \circ (a \cdot u') \in W^{1,1}(\mathbb{R})$, which satisfies (7.1.1),(7.1.2).

The following will be assumed:

$(H_1)$ $\phi$ is an increasing homeomorphism with $\phi(0) = 0$ and $\phi(\mathbb{R}) = \mathbb{R}$ such that

$$\left| \phi^{-1}(x) \right| \leq \phi^{-1}(|x|); \tag{7.2.1}$$

$(H_2)$ $a \in C(\mathbb{R})$ with $a(t) > 0$, $\forall t \in \mathbb{R}$, such that $\lim_{|t| \to +\infty} \frac{1}{a(t)} \in \mathbb{R}$, and

$$\int_{-\infty}^{+\infty} \frac{ds}{a(s)} < +\infty. \tag{7.2.2}$$

The next result states the relation between the boundary value problem (7.1.1),(7.1.2) and the correspondent integral form.

**Lemma 7.2.1.** *Suppose that $f$ is an $L^1$-Carathéodory and assumptions $(H_1)$, $(H_2)$ hold. Then $u \in X$ is a solution of problem (7.1.1),(7.1.2) if and only if*

$$u(t) = A + \int_{-\infty}^{t} \frac{1}{a(s)} \phi^{-1} \left( \tau_u + \int_{-\infty}^{s} f(r, u(r), u'(r)) \, dr \right) ds, \tag{7.2.3}$$

*where $\tau_u$ is the unique solution of the equation*

$$\int_{-\infty}^{+\infty} \frac{1}{a(s)} \phi^{-1} \left( \tau_u + \int_{-\infty}^{s} f(r, u(r), u'(r)) \, dr \right) ds = B - A. \tag{7.2.4}$$

*Moreover,*

$$\tau_u \in [y_{1u}, y_{3u}] \tag{7.2.5}$$

*with*

$$y_{1u} = -\int_{-\infty}^{+\infty} |f(r, u(r), u'(r))| \, dr \tag{7.2.6}$$

*and $y_{3u}$ to be defined forward.*

**Proof.** Let $u$ be a solution of problem (7.1.1),(7.1.2). So, for some constant $\tau_u \in \mathbb{R}$,

$$\phi(a(t)u'(t)) = \tau_u + \int_{-\infty}^{t} f(r, u(r), u'(r)) dr, \quad \text{for } t \in \mathbb{R},$$

and

$$u'(t) = \frac{1}{a(t)} \phi^{-1} \left( \tau_u + \int_{-\infty}^{t} f(r, u(r), u'(r)) \, dr \right).$$

By (7.1.2),

$$u(t) = A + \int_{-\infty}^{t} \frac{1}{a(s)} \phi^{-1} \left( \tau_u + \int_{-\infty}^{s} f(r, u(r), u'(r)) \, dr \right) ds$$

and

$$A + \int_{-\infty}^{+\infty} \frac{1}{a(s)} \phi^{-1} \left( \tau_u + \int_{-\infty}^{s} f(r, u(r), u'(r)) \, dr \right) ds = B.$$

To show that $\tau_u$ is the unique solution of (7.2.4), consider the function

$$F(y_u) := \int_{-\infty}^{+\infty} \frac{1}{a(s)} \phi^{-1} \left( y_u + \int_{-\infty}^{s} f(r, u(r), u'(r)) \, dr \right) ds$$

and remark that $F(y)$ is strictly increasing on $\mathbb{R}$,

$$\lim_{y_u \to -\infty} F(y_u) = -\infty \quad \text{and} \quad \lim_{y_u \to +\infty} F(y_u) = +\infty. \tag{7.2.7}$$

Moreover,

$$F(y_{1u}) = \int_{-\infty}^{+\infty} \frac{1}{a(s)} \phi^{-1} \left( y_{1u} + \int_{-\infty}^{s} f(r, u(r), u'(r)) \, dr \right) ds \leq 0,$$

for $y_{1u}$ given by (7.2.6), and, for

$$y_{2u} = \int_{-\infty}^{+\infty} |f(r, u(r), u'(r))| \, dr, \tag{7.2.8}$$

we have

$$F(y_{2u}) = \int_{-\infty}^{+\infty} \frac{1}{a(s)} \phi^{-1} \left( y_{2u} + \int_{-\infty}^{s} f(r, u(r), u'(r)) \, dr \right) ds \geq 0.$$

By (7.2.7), there are $k > 0$ and

$$y_{3u} := y_{2u} + k \qquad (7.2.9)$$

such that

$$F(y_{3u}) > F(y_{2u}) + B - A.$$

Therefore, the equation $F(y) = B - A$ has a unique solution $\tau_u$, and $\tau_u \in [y_{1u}, y_{3u}]$.

If $u(t)$ verifies (7.2.3) and (9.2.5), then by standard arguments, it can be shown that $u(t)$ is a solution of problem (7.1.1),(7.1.2). $\qquad \square$

To overcome the lack of compactness of set $X$, we apply the following compactness criterion, suggested in [51].

**Lemma 7.2.2.** *A set $M \subset X$ is compact if the following conditions hold:*

(1) *$M$ is uniformly bounded in $X$;*
(2) *the functions belonging to $M$ are equicontinuous on any compact interval of $\mathbb{R}$;*
(3) *the functions from $M$ are equiconvergent at $\pm\infty$, that is, given $\epsilon > 0$, there exists $T(\epsilon) > 0$ such that $|f(t) - f(\pm\infty)| < \epsilon$ and $|f'(t) - f'(\pm\infty)| < \epsilon$, for all $|t| > T(\epsilon)$ and $f \in M$.*

## 7.3. Existence of heteroclinics

The main theorem gives sufficient conditions for the existence of heteroclinic solutions of problem (7.1.1),(7.1.2), without asymptotic, growth or extra conditions on the homeomorphism $\phi$ or on the nonlinearity $f$.

**Theorem 7.3.1.** *Assume that $f : \mathbb{R}^3 \to \mathbb{R}$ is an $L^1$-Carathéodory function and assumptions $(H_1)$, $(H_2)$ hold. Then problem (7.1.1),(7.1.2) has at least a solution $u \in X$, that is, a heteroclinic solution of (7.1.1).*

**Proof.** Define the operator $T : X \to X$ given by

$$Tu(t) = A + \int_{-\infty}^{t} \frac{1}{a(s)} \phi^{-1} \left( \tau_u + \int_{-\infty}^{s} f(r, u(r), u'(r)) \, dr \right) ds$$

with $\tau_u$ the unique solution of (7.2.4).

To prove the theorem by Lemma 7.2.1, it is enough to show that $T$ has a fixed point.

**Claim 1.** $T : X \to X$ *is well defined.*

Let $u \in X$. So, there is $\rho > 0$, such that $\|u\|_X < \rho$.

As $f$ is an $L^1$-Carathéodory function then there exists a positive function $\varphi_\rho \in L^1(\mathbb{R})$ such that $|f(t, u(t), u'(t))| \leq \varphi_\rho(t)$, a.e. $t \in \mathbb{R}$, and

$$\int_{-\infty}^{t} |f(r, u(r), u'(r))| \, dr \leq \int_{-\infty}^{+\infty} |f(r, u(r), u'(r))| \, dr$$

$$\leq \int_{-\infty}^{+\infty} \varphi_\rho(t) dt < +\infty. \qquad (7.3.1)$$

So, by $(H_1)$, $(H_2)$, $Tu$ is continuous on $\mathbb{R}$.
In the same way, it is clear that

$$(Tu)'(t) = \frac{1}{a(t)} \phi^{-1} \left( \tau_u + \int_{-\infty}^{t} f(r, u(r), u'(r)) \, dr \right)$$

is continuous on $\mathbb{R}$. Therefore $Tu \in C^1(\mathbb{R})$.
Moreover,

$$\lim_{t \to -\infty} Tu(t) = \lim_{t \to -\infty} A + \int_{-\infty}^{t} \frac{1}{a(s)} \phi^{-1}$$

$$\times \left( \tau_u + \int_{-\infty}^{s} f(r, u(r), u'(r)) \, dr \right) ds = A,$$

by (7.2.4),

$$\lim_{t \to +\infty} Tu(t) = \lim_{t \to +\infty} A + \int_{-\infty}^{t} \frac{1}{a(s)} \phi^{-1}$$

$$\times \left( \tau_u + \int_{-\infty}^{s} f(r, u(r), u'(r)) \, dr \right) ds = B,$$

and, by $(H_1)$, $(H_2)$ and (7.3.1),

$$\lim_{t \to \pm\infty} (Tu)'(t) = \lim_{t \to \pm\infty} \frac{1}{a(t)} \phi^{-1}$$

$$\times \left( \tau_u + \int_{-\infty}^{t} f(r, u(r), u'(r)) \, dr \right) \in \mathbb{R}.$$

Therefore $Tu \in X$.

**Claim 2.** *T is compact.*

Let $B \subset X$ be a bounded subset and $u \in B$. Then there is $\rho_1 > 0$ such that $\|u\|_X < \rho_1$, that is, $\|u\|_\infty < \rho_1$ and $\|u'\|_\infty < \rho_1$.

To apply Lemma 7.2.2, we follow several steps:

**Step 2.1.** *TB is uniformly bounded, for B any bounded set in X.*

By (7.2.1), (7.2.5) and $(H_1)$, we have

$$\|Tu\|_\infty = \sup_{t\in\mathbb{R}} \left| A + \int_{-\infty}^t \frac{1}{a(s)} \phi^{-1}\left(\tau_u + \int_{-\infty}^s f\left(r, u(r), u'(r)\right) dr\right) ds \right|$$

$$\leq \sup_{t\in\mathbb{R}} |A| + \int_{-\infty}^t \frac{1}{a(s)} \left| \phi^{-1}\left(\tau_u + \int_{-\infty}^s f\left(r, u(r), u'(r)\right) dr\right) \right| ds$$

$$\leq \sup_{t\in\mathbb{R}} |A| + \int_{-\infty}^t \frac{1}{a(s)} \phi^{-1}\left(|\tau_u| + \int_{-\infty}^s |f\left(r, u(r), u'(r)\right)| dr\right) ds$$

$$\leq |A| + \int_{-\infty}^{+\infty} \frac{1}{a(s)} \phi^{-1}\left(|\tau_u| + \int_{-\infty}^s \varphi_{\rho_1}(r) dr\right) ds$$

$$\leq |A| + \phi^{-1}\left(2 \int_{-\infty}^{+\infty} \varphi_{\rho_1}(r) dr + k\right) \int_{-\infty}^{+\infty} \frac{1}{a(s)} ds < +\infty,$$

and

$$\left\|(Tu)'\right\|_\infty = \sup_{t\in\mathbb{R}} \left| \frac{1}{a(t)} \phi^{-1}\left(\tau_u + \int_{-\infty}^t f\left(r, u(r), u'(r)\right) dr\right) \right|$$

$$\leq \sup_{t\in\mathbb{R}} \frac{1}{a(t)} \phi^{-1}\left(|\tau_u| + \int_{-\infty}^t |f\left(r, u(r), u'(r)\right)| dr\right)$$

$$\leq \sup_{t\in\mathbb{R}} \frac{1}{a(t)} \phi^{-1}\left(|\tau_u| + \int_{-\infty}^{+\infty} \varphi_{\rho_1}(r) dr\right)$$

$$\leq \phi^{-1}\left(2 \int_{-\infty}^{+\infty} \varphi_{\rho_1}(r) dr + k\right) \sup_{t\in\mathbb{R}} \frac{1}{a(t)} < +\infty.$$

Therefore, $TB$ is uniformly bounded in $X$.

**Step 2.2.** *TB is equicontinuous on X.*

For $L > 0$ consider $t_1, t_2 \in [-L, L]$. Assume, without loss of generality, that $t_1 < t_2$.

Then, by (7.2.1), (7.2.5) and $(H_1)$,

$$
\begin{aligned}
|Tu(t_1) - Tu(t_2)| &= \left| \int_{-\infty}^{t_1} \frac{1}{a(s)} \phi^{-1} \left( \tau_u + \int_{-\infty}^{s} f(r, u(r), u'(r)) \, dr \right) ds \right. \\
&\qquad \left. - \int_{-\infty}^{t_2} \frac{1}{a(s)} \phi^{-1} \left( \tau_u + \int_{-\infty}^{s} f(r, u(r), u'(r)) \, dr \right) ds \right| \\
&= \left| \int_{t_1}^{t_2} \frac{1}{a(s)} \phi^{-1} \left( \tau_u + \int_{-\infty}^{s} f(r, u(r), u'(r)) \, dr \right) ds \right| \\
&\leq \int_{t_1}^{t_2} \frac{1}{a(s)} \phi^{-1} \left( |\tau_u| + \int_{-\infty}^{s} |f(r, u(r), u'(r))| \, dr \right) ds \\
&\leq \phi^{-1} \left( 2 \int_{-\infty}^{+\infty} \varphi_{\rho_1}(r) dr + k \right) \int_{t_1}^{t_2} \frac{1}{a(s)} ds \\
&\longrightarrow 0, \text{ uniformly in } u \in B, \text{ as } t_1 \to t_2,
\end{aligned}
$$

and

$$
\begin{aligned}
|(Tu)'(t_1) - (Tu)'(t_2)| &= \left| \frac{1}{a(t_1)} \phi^{-1} \left( \tau_u + \int_{-\infty}^{t_1} f(r, u(r), u'(r)) \, dr \right) \right. \\
&\qquad \left. - \frac{1}{a(t_2)} \phi^{-1} \left( \tau_u + \int_{-\infty}^{t_2} f(r, u(r), u'(r)) \, dr \right) \right| \\
&\leq \phi^{-1} \left( 2 \int_{-\infty}^{+\infty} \varphi_{\rho_1}(r) dr + k \right) \left( \frac{1}{a(t_1)} - \frac{1}{a(t_2)} \right) \\
&\longrightarrow 0, \text{ uniformly in } u \in B, \text{ as } t_1 \to t_2.
\end{aligned}
$$

So, $TB$ is equicontinuous on $X$.

**Step 2.3.** $TB$ *is equiconvergent at* $\pm\infty$.
Let $u \in B$. Then, as previously,

$$
\begin{aligned}
&\left| Tu(t) - \lim_{t \to -\infty} (Tu(t)) \right| \\
&= \left| \int_{-\infty}^{t} \frac{1}{a(s)} \phi^{-1} \left( \tau_u + \int_{-\infty}^{s} f(r, u(r), u'(r)) \, dr \right) ds \right| \\
&\leq \phi^{-1} \left( 2 \int_{-\infty}^{+\infty} \varphi_{\rho_1}(r) dr \right) \int_{-\infty}^{t} \frac{1}{a(s)} ds \\
&\longrightarrow 0, \text{ uniformly in } u \in B, \text{ as } t \to -\infty,
\end{aligned}
$$

and

$$\left| Tu(t) - \lim_{t \to +\infty} (Tu(t)) \right|$$

$$= \left| \int_{-\infty}^{t} \frac{1}{a(s)} \phi^{-1} \left( \tau_u + \int_{-\infty}^{s} f(r, u(r), u'(r)) \, dr \right) ds \right.$$

$$\left. - \int_{-\infty}^{+\infty} \frac{1}{a(s)} \phi^{-1} \left( \tau_u + \int_{-\infty}^{s} f(r, u(r), u'(r)) \, dr \right) ds \right|$$

$$= \left| \int_{t}^{+\infty} \frac{1}{a(s)} \phi^{-1} \left( \tau_u + \int_{-\infty}^{s} f(r, u(r), u'(r)) \, dr \right) ds \right|$$

$$\leq \phi^{-1} \left( 2 \int_{-\infty}^{+\infty} \varphi_{\rho_1}(r) dr + k \right) \int_{t}^{+\infty} \frac{1}{a(s)} ds$$

$$\longrightarrow 0, \text{ uniformly in } u \in B, \text{ as } t \to +\infty.$$

Moreover,

$$\left| (Tu)'(t) - \lim_{t \to -\infty} (Tu)'(t) \right|$$

$$= \left| \frac{1}{a(t)} \phi^{-1} \left( \tau_u + \int_{-\infty}^{t} f(r, u(r), u'(r)) \, dr \right) \right.$$

$$\left. - \lim_{t \to -\infty} \frac{1}{a(t)} \phi^{-1}(\tau_u) \right|$$

$$\leq \left| \frac{1}{a(t)} \phi^{-1} \left( \tau_u + \int_{-\infty}^{t} \varphi_{\rho_1}(r) dr \right) - \lim_{t \to -\infty} \frac{1}{a(t)} \phi^{-1}(\tau_u) \right|$$

$$\longrightarrow 0, \text{ uniformly in } u \in B, \text{ as } t \to -\infty,$$

and

$$\left| (Tu)'(t) - \lim_{t \to +\infty} (Tu)'(t) \right|$$

$$= \left| \frac{1}{a(t)} \phi^{-1} \left( \tau_u + \int_{-\infty}^{t} f(r, u(r), u'(r)) \, dr \right) \right.$$

$$\left. - \lim_{t \to +\infty} \frac{1}{a(t)} \phi^{-1} \left( \tau_u + \int_{-\infty}^{+\infty} f(r, u(r), u'(r)) \, dr \right) \right|$$

$$\leq \phi^{-1} \left( 2 \int_{-\infty}^{+\infty} \varphi_{\rho_1}(r) dr + k \right) \left| \frac{1}{a(t)} - \lim_{t \to +\infty} \frac{1}{a(t)} \right|$$

$$\longrightarrow 0, \text{ uniformly in } u \in B, \text{ as } t \to +\infty.$$

Therefore, $TB$ is equiconvergent at $\pm\infty$, and, by Lemma 7.2.2, $T$ is compact.

**Claim 3.** $TD \subset D$ *for* $D \subset X$ *a closed and bounded set.*

Consider $D \subset X$ defined by

$$D = \{x \in X : \|x\|_X \le \rho_2\}$$

with $\rho_2$ given by

$$\rho_2 > \max\left\{\rho_1, \ |A| + K\int_{-\infty}^{+\infty}\frac{1}{a(s)}ds, \ K\sup_{t\in\mathbb{R}}\frac{1}{a(t)}\right\},$$

where

$$K := \phi^{-1}\left(2\int_{-\infty}^{+\infty}\varphi_{\rho_1}(r)dr + k\right).$$

Applying the same technique as in Step 2.1, we have

$$\|Tu\|_\infty = \sup_{t\in\mathbb{R}}|Tu(t)|$$

$$\le |A| + \int_{-\infty}^{+\infty}\frac{1}{a(s)}\phi^{-1}\left(|\tau_u| + \int_{-\infty}^{s}\varphi_{\rho_1}(r)dr\right)ds$$

$$\le |A| + \phi^{-1}\left(2\int_{-\infty}^{+\infty}\varphi_{\rho_1}(r)dr + k\right)\int_{-\infty}^{+\infty}\frac{1}{a(s)}ds < \rho_2,$$

and

$$\|(Tu)'\|_\infty = \sup_{t\in\mathbb{R}}|(Tu)'(t)|$$

$$\le \sup_{t\in\mathbb{R}}\frac{1}{a(t)}\phi^{-1}\left(|\tau_u| + \int_{-\infty}^{t}|f(r,u(r),u'(r))|\,dr\right)$$

$$\le \phi^{-1}\left(2\int_{-\infty}^{+\infty}\varphi_{\rho_1}(r)dr + k\right)\sup_{t\in\mathbb{R}}\frac{1}{a(t)} < \rho_2.$$

So, $TD \subset D$ and, by Theorem 1.2.6, $T$ has at least one fixed point $u \in X$, which, by Lemma.7.2.1, is a heteroclinic solution of (7.1.1). $\qquad\square$

## 7.4. Example

Consider the second-order differential equation

$$[((t^2+1)\,u'(t))^3]' = \frac{[(u(t))^2-1][(u'(t))^6+1]}{1+t^2}, \quad \text{a.e. } t \in \mathbb{R}, \qquad (7.4.1)$$

and the boundary conditions

$$u(-\infty) = -1, \quad u(+\infty) = 1.$$

The above problem is a particular case of problem (7.1.1),(7.1.2), with

$$\phi(w) = w^3,$$

$$a(t) = 1 + t^2,$$

$$f(t, x, y) = \frac{\left(x^2 - 1\right)\left(y^6 + 1\right)}{1 + t^2},$$

$$A = -1, \ B = 1.$$

All assumptions of Theorem 7.3.1 are satisfied, namely $f$ is an $L^1$-Carathéodory function with

$$\varphi_\rho(t) = \frac{\left(\rho^2 + 1\right)\left(\rho^6 + 1\right)}{1 + t^2},$$

and therefore, there is a heteroclinic solution of (7.4.1) linking the two equilibrium points $-1$ and $1$.

We point out that, as far as we know, the existence of heteroclinic solutions for (7.4.1) was not covered by the existent literature, namely, because the nonlinearity $f$ does not verify the asymptotic conditions in [9, 43, 112], or the growth assumptions of [102], for example.

# Chapter 8

# Heteroclinic Solutions for Semi-linear Problems (ii)

## 8.1. Introduction

This chapter considers the second-order nonlinear discontinuous equation in the real line,

$$(a(t)\phi(u'(t)))' = f(t, u(t), u'(t)), \quad \text{a.e. } t \in \mathbb{R}, \tag{8.1.1}$$

where $\phi$ is an increasing homeomorphism with $\phi(0) = 0$ and $\phi(\mathbb{R}) = \mathbb{R}$, $a \in C(\mathbb{R}, \mathbb{R} \setminus \{0\}) \cap C^1(\mathbb{R}, \mathbb{R})$ with $a(t) > 0$, or $a(t) < 0$, for $t \in \mathbb{R}$, and $f : \mathbb{R}^3 \to \mathbb{R}$ an $L^1$-Carathéodory function.

We are looking for heteroclinic solutions, that is, nontrivial solutions of (8.1.1) such that

$$u(-\infty) := \lim_{t \to -\infty} u(t) = \nu^-, u(+\infty) := \lim_{t \to +\infty} u(t) = \nu^+, \tag{8.1.2}$$

with $\nu^-, \nu^+ \in \mathbb{R}$ such that $\nu^- < \nu^+$.

In [53], the lower and upper solutions method is applied to study the equation

$$(a(x(t))\phi(x'(t)))' = f(t, x(t), x'(t)), \quad \text{a.e. } t,$$

where $a : \mathbb{R} \to \mathbb{R}$ is a positive continuous function, and $f : \mathbb{R}^3 \to \mathbb{R}$ is a Carathéodory function verifying, in short, the following assumptions on $f$:

- there exist a constant $H > 0$, a continuous function $\theta : \mathbb{R}^+ \to \mathbb{R}^+$ and a function $\lambda \in L^q([-L, L])$, with $1 \leq q \leq \infty$, such that

$$|f(t, x, y)| \leq \lambda(t)\theta(a(t, x)|\phi(y)|), \quad \text{for a.e. } |t| \leq L, \qquad (8.1.3)$$

  every $x \in \mathcal{I} : = [\inf_{t \in \mathbb{R}} \alpha(t), \sup_{t \in \mathbb{R}} \beta(t)]$, and $|y| \geq H$;
- for every $C > 0$, there exist functions $\eta_C \in L^1(\mathbb{R})$, $K_C \in W^{1,1}_{\text{loc}}([0, +\infty))$, null in $[0, L]$ and positive in $[L, +\infty)$, and $N_C(t) \in L^1(\mathbb{R})$ such that

$$f(t, x, y) \leq -K'_C(t)\phi(|y|),$$

$$f(-t, x, y) \geq K'_C(t)\phi(|y|), \quad \text{for a.e. } t \geq L, \text{ every } x \in \mathcal{I}, |y| \leq N_C(t),$$

$$|f(t, x, y)| \leq \eta_C(t) \text{ if } x \in \mathcal{I}, |y| \leq N_C(t)$$

$$+ |\alpha'(t)| + |\beta'(t)|, \quad \text{for a.e. } t \in \mathbb{R}. \qquad (8.1.4)$$

In [112], the author considers

$$(a(t, x(t))\phi(x'(t)))' = f(t, x(t), x'(t)), \quad \text{a.e. } t,$$

$$x(-\infty) = \nu^-, \ x(+\infty) = \nu^+,$$

with $\phi$ a general increasing homeomorphism on $\mathbb{R}$, $a : \mathbb{R}^2 \to \mathbb{R}$ a positive continuous function and $f : \mathbb{R}^3 \to \mathbb{R}$ a Carathéodory function verifying, in short the following:

- $\phi$ has a definite growth at infinity (sublinear, linear or superlinear);
- $f(t, \nu^-, 0) \leq 0 \leq f(t, \nu^+, 0)$, for a.e. $t \in \mathbb{R}$;
- there exist constants $L, H > 0$, a continuous function $\theta : \mathbb{R}^+ \to \mathbb{R}^+$ and a function $\lambda \in L^q([-L, L])$, with $1 \leq q \leq \infty$, such that

$$|f(t, x, y)| \leq \lambda(t)\theta(a(t, x)|\phi(y)|), \quad \text{for a.e. } |t| \leq L,$$

  every $x \in [\nu^-, \nu^+]$, and $|y| \geq H$;
- for every $C > 0$, there exist functions $\eta_C \in L^1(\mathbb{R})$, $\Lambda_C \in L^1_{\text{loc}}([0, +\infty))$, null in $[0, L]$ and positive in $[L, +\infty)$, and $N_C(t) \in L^1(\mathbb{R})$ such that

$$f(t, x, y) \leq -\Lambda_C(t)(|y|),$$

$$f(-t, x, y) \geq \Lambda_C(t)(|y|), \quad \text{for a.e. } t \geq L,$$

for every $x \in [\nu^-, \nu^+]$, $|y| \leq N_C(t)$,

$$|f(t, x, y)| \leq \eta_C(t) \quad \text{if } x \in [\nu^-, \nu^+], \quad |y| \leq N_C(t), \quad \text{for a.e. } t \in \mathbb{R}.$$

Motivated by these two works, we consider Eq. (8.1.1) where the function $a(t)$ must have a definite sign, but can be positive or negative. This information is important in some applications to traveling wave solutions for reaction–diffusion equations: diffusion phenomena if $a(t)$ is positive, diffusion–aggregation processes if $a(t)$ changes sign (see, for example, [25, 61, 70]).

We point out that, in this work, the existence of heteroclinic solutions for (8.1.1) is obtained without asymptotic growth or other extra assumptions on the nonlinearities $\phi$ and $f$, applying similar techniques suggested in [71, 72]. On the other hand, our method remains valid for $\phi \equiv I$, that is, for equation

$$\left( a(t) u'(t) \right)' = f\left( t, u(t), u'(t) \right), \quad \text{a.e. } t \in \mathbb{R}.$$

The study of boundary value problems on the whole real line, and the existence of homoclinic or heteroclinic solutions, had an increasing interest in the recent years due to the applications to non-Newtonian fluids theory, diffusion of flows in porous media, nonlinear elasticity (see, for instance, [9, 25, 43, 85, 101, 111, 135, 138, 140] and the references therein). In particular, heteroclinic connections are related to processes in which the variable transits from an unstable equilibrium to a stable one (see, for example, [14, 23, 37, 42, 59, 61, 70, 108, 123, 127]). In this sense, heteroclinic solutions are often referred as transitional solutions.

## 8.2. Auxiliary results

The functional set is defined as

$$X = \left\{ u \in C^1(\mathbb{R}) : \lim_{|t| \to +\infty} u^{(i)}(t) \in \mathbb{R}, \quad i = 0, 1 \right\}$$

with the norm

$$\|x\|_X = \max \left\{ \|x\|_\infty, \|x'\|_\infty \right\}, \quad \text{where } \|y\|_\infty := \sup_{t \in \mathbb{R}} |y(t)|.$$

It can be proved, by standard arguments, that $(X, \|\cdot\|_X)$ is a Banach space.

By a solution of problem (8.1.1),(8.1.2), we consider a function $u \in X$ such that $a \cdot (\phi \circ u') \in W^{1,1}(\mathbb{R})$, satisfying (8.1.1),(8.1.2).

The following assumptions will be considered forward:

$(A_1)$ $\phi$ is an increasing homeomorphism with $\phi(0) = 0$ and $\phi(\mathbb{R}) = \mathbb{R}$ such that

$$\left|\phi^{-1}(w)\right| \leq \phi^{-1}(|w|); \qquad (8.2.1)$$

$(A_2)$ $a \in C(\mathbb{R}, \mathbb{R} \backslash \{0\})$ with $a(t) > 0$, or $a(t) < 0$, $\forall t \in \mathbb{R}$, such that

$$\lim_{|t| \to +\infty} |a(t)| = +\infty$$

and

$$\int_{-\infty}^{+\infty} \phi^{-1}\left( \frac{2 \int_{-\infty}^{+\infty} \varphi_\rho(r) dr}{|a(s)|} \right) ds < +\infty. \qquad (8.2.2)$$

The solvability of the integral equation associated to the problem (8.1.1),(8.1.2) is studied in the next lemma.

**Lemma 8.2.1.** *Consider that $f$ is an $L^1$-Carathéodory and assumptions $(A_1)$, $(A_2)$ hold. Then $u \in X$ is a solution of problem (8.1.1),(8.1.2) if and only if*

$$u(t) = \nu^- + \int_{-\infty}^{t} \phi^{-1}\left( \frac{\tau_u + \int_{-\infty}^{s} f\left(r, u(r), u'(r)\right) dr}{a(s)} \right) ds \qquad (8.2.3)$$

*with $\tau_u$ the unique solution of the equation*

$$\int_{-\infty}^{+\infty} \phi^{-1}\left( \frac{\tau_u + \int_{-\infty}^{s} f\left(r, u(r), u'(r)\right) dr}{a(s)} \right) ds = \nu^+ - \nu^-. \qquad (8.2.4)$$

*Moreover,*

$$\tau_u \in [w_1, w_2], \quad if \ a(t) > 0, \ \forall t \in \mathbb{R}, \qquad (8.2.5)$$

*or*

$$\tau_u \in [w_2, w_1], \quad if \ a(t) < 0, \ \forall t \in \mathbb{R}, \qquad (8.2.6)$$

*with*

$$w_1 := - \int_{-\infty}^{+\infty} |f(r, u(r), u'(r))| \, dr \tag{8.2.7}$$

*and*

$$w_2 := \int_{-\infty}^{+\infty} |f(r, u(r), u'(r))| \, dr. \tag{8.2.8}$$

**Proof.** For $u$ a solution of problem (8.1.1),(8.1.2), there is a constant $\tau_u \in \mathbb{R}$, such that

$$a(t) \, \phi(u'(t)) = \tau_u + \int_{-\infty}^{t} f(r, u(r), u'(r)) \, dr, \quad \text{for } t \in \mathbb{R},$$

and

$$u'(t) = \phi^{-1} \left( \frac{\tau_u + \int_{-\infty}^{s} f(r, u(r), u'(r)) \, dr}{a(t)} \right).$$

By (8.1.2),

$$u(t) = \nu^- + \int_{-\infty}^{t} \phi^{-1} \left( \frac{\tau_u + \int_{-\infty}^{s} f(r, u(r), u'(r)) \, dr}{a(s)} \right) ds$$

and

$$\nu^- + \int_{-\infty}^{+\infty} \phi^{-1} \left( \frac{\tau_u + \int_{-\infty}^{s} f(r, u(r), u'(r)) \, dr}{a(s)} \right) ds = \nu^+.$$

To see that $\tau_u$ is the unique solution of (8.2.4), define the function

$$F(y) := \int_{-\infty}^{+\infty} \phi^{-1} \left( \frac{y + \int_{-\infty}^{s} f(r, u(r), u'(r)) \, dr}{a(s)} \right) ds.$$

As $f$ is an $L^1$-Carathéodory function, by $(A_1)$ and $(A_2)$, $F$ is well defined. If $a(t) > 0$, $\forall t \in \mathbb{R}$, then $F(y)$ is strictly increasing in $\mathbb{R}$, and

$$\lim_{y \to +\infty} F(y) = \int_{-\infty}^{+\infty} \phi^{-1}(+\infty) \, ds = +\infty,$$

$$\lim_{y \to -\infty} F(y) = \int_{-\infty}^{+\infty} \phi^{-1}(-\infty) \, ds = -\infty.$$

Otherwise, if $a(t) < 0$, $\forall t \in \mathbb{R}$, then $F(y)$ is strictly decreasing in $\mathbb{R}$, and

$$\lim_{y \to +\infty} F(y) = \int_{-\infty}^{+\infty} \phi^{-1}(-\infty)\, ds = -\infty,$$

$$\lim_{y \to -\infty} F(y) = \int_{-\infty}^{+\infty} \phi^{-1}(+\infty)\, ds = +\infty.$$

Therefore, the equation $F(y) = \nu^- - \nu^+$ has a unique solution $\tau_u$.

Moreover, $F(w_1)$ and $F(w_2)$ have opposite signs. For example, in the case $a(t) > 0$, $\forall t \in \mathbb{R}$, we have

$$F(w_1) = \int_{-\infty}^{+\infty} \phi^{-1}\left(\frac{w_1 + \int_{-\infty}^{s} f\left(r, u(r), u'(r)\right) dr}{a(s)}\right) ds \leq 0,$$

for $w_1$ given by (8.2.7), and

$$F(w_2) = \int_{-\infty}^{+\infty} \phi^{-1}\left(\frac{w_2 + \int_{-\infty}^{s} f\left(r, u(r), u'(r)\right) dr}{a(s)}\right) ds \geq 0,$$

for $w_2$ given by (8.2.8). So $\tau_u \in [w_1, w_2]$, if $a(t) > 0$, $\forall t \in \mathbb{R}$, and $\tau_u \in [w_2, w_1]$, if $a(t) < 0$, $\forall t \in \mathbb{R}$. □

## 8.3. Existence of heteroclinics solutions

The main result presents sufficient conditions for the existence of heteroclinic solutions of problem (8.1.1),(8.1.2) without the usual asymptotic or growth assumptions on $\phi$ or on $f$.

**Theorem 8.3.1.** *Suppose that $f : \mathbb{R}^3 \to \mathbb{R}$ is an $L^1$-Carathéodory function and hypothesis $(A_1)$, $(A_2)$ hold. Then problem (8.1.1),(8.1.2) has at least a solution $u \in X$, that is, there is a heteroclinic solution for (8.1.1).*

**Proof.** Define the operator $T : X \to X$ by

$$Tu(t) = \nu^- + \int_{-\infty}^{t} \phi^{-1}\left(\frac{\tau_u + \int_{-\infty}^{s} f\left(r, u(r), u'(r)\right) dr}{a(s)}\right) ds,$$

where $\tau_u$ is the unique solution of (8.2.4).

From Lemma 8.2.1, it is enough to prove that $T$ has a fixed point. For clearance, we use several steps.

**Step 1.** $T : X \to X$ *is well defined.*
For each $u \in X$, there is $\rho > 0$ such that $\|u\|_X < \rho$, and, as $f$ is an $L^1$-Carathéodory function, a positive function $\varphi_\rho \in L^1(\mathbb{R})$ such that

$$|f(t, u(t), u'(t))| \leq \varphi_\rho(t), \quad \text{a.e. } t \in \mathbb{R},$$

and

$$\int_{-\infty}^{t} f(r, u(r), u'(r))\, dr \leq \int_{-\infty}^{+\infty} \varphi_\rho(t) dt < +\infty. \tag{8.3.1}$$

By $(A_1)$, $(A_2)$, $Tu$ is continuous on $\mathbb{R}$.
For the derivative of the operator,

$$(Tu)'(t) = \phi^{-1}\left(\frac{\tau_u + \int_{-\infty}^{t} f(r, u(r), u'(r))\, dr}{a(t)}\right)$$

is continuous on $\mathbb{R}$, and, therefore, $Tu \in C^1(\mathbb{R})$.

$$\lim_{t \to -\infty} Tu(t) = \lim_{t \to -\infty} \nu^- + \int_{-\infty}^{t} \phi^{-1}\left(\frac{\tau_u + \int_{-\infty}^{s} f(r, u(r), u'(r))dr}{a(s)}\right) ds = \nu^-,$$

by (8.2.4),

$$\lim_{t \to +\infty} Tu(t) = \lim_{t \to +\infty} \nu^- + \int_{-\infty}^{t} \phi^{-1}\left(\frac{\tau_u + \int_{-\infty}^{s} f(r, u(r), u'(r))dr}{a(s)}\right) ds = \nu^+.$$

From $(A_1)$, $(A_2)$, (8.2.5), or (8.2.6), and (8.3.1),

$$\lim_{t \to -\infty} (Tu)'(t) = \lim_{t \to -\infty} \phi^{-1}\left(\frac{\tau_u + \int_{-\infty}^{t} f(r, u(r), u'(r))\, dr}{a(t)}\right)$$

$$= \phi^{-1}\left(\frac{\tau_u}{\lim_{t \to -\infty} a(t)}\right) = 0$$

and

$$\lim_{t \to +\infty} (Tu)'(t) = \phi^{-1} \left( \frac{\tau_u + \int_{-\infty}^{+\infty} f(r, u(r), u'(r)) \, dr}{\lim_{t \to +\infty} a(t)} \right) = 0.$$

So, $Tu \in X$.

**Step 2.** *T is compact.*

Consider a bounded subset $B \subset X$, $u \in B$, and $\rho_0 > 0$ such that $\|u\|_X < \rho_0$. Therefore, $\|u\|_\infty < \rho_0$ and $\|u'\|_\infty < \rho_0$.

To verify the assumptions of Lemma 7.2.2, we consider some claims.

**Claim 2.1.** *TB is uniformly bounded, for B a bounded set in X.*

By (8.2.1), (8.2.5), or (8.2.6), and $(A_1)$, we have

$$\|Tu\|_\infty = \sup_{t \in \mathbb{R}} \left| \nu^- + \int_{-\infty}^t \phi^{-1} \left( \frac{\tau_u + \int_{-\infty}^s f(r, u(r), u'(r)) dr}{a(s)} \right) ds \right|$$

$$\leq \sup_{t \in \mathbb{R}} |\nu^-| + \int_{-\infty}^t \phi^{-1} \left( \left| \frac{\tau_u + \int_{-\infty}^s f(r, u(r), u'(r)) dr}{a(s)} \right| \right) ds$$

$$\leq \sup_{t \in \mathbb{R}} |\nu^-| + \int_{-\infty}^t \phi^{-1} \left( \frac{|\tau_u| + \int_{-\infty}^s |f(r, u(r), u'(r))|}{|a(s)|} dr \right) ds$$

$$\leq |\nu^-| + \int_{-\infty}^{+\infty} \phi^{-1} \left( \frac{|\tau_u| + \int_{-\infty}^s \varphi_{\rho_0}(r) dr}{|a(s)|} \right) ds$$

$$\leq |\nu^-| + \int_{-\infty}^{+\infty} \phi^{-1} \left( \frac{2 \int_{-\infty}^{+\infty} \varphi_{\rho_0}(r) dr}{|a(s)|} \right) ds < +\infty,$$

and, by $(A_2)$,

$$\|(Tu)'\|_\infty = \sup_{t \in \mathbb{R}} \left| \phi^{-1} \left( \frac{\tau_u + \int_{-\infty}^t f(r, u(r), u'(r)) \, dr}{a(t)} \right) \right|$$

$$\leq \sup_{t \in \mathbb{R}} \phi^{-1} \left( \frac{|\tau_u| + \int_{-\infty}^t |f(r, u(r), u'(r))| dr}{|a(t)|} \right)$$

$$\leq \sup_{t \in \mathbb{R}} \phi^{-1} \left( \frac{|\tau_u| + \int_{-\infty}^{+\infty} \varphi_{\rho_0}(r) dr}{|a(t)|} \right)$$

$$\leq \sup_{t \in \mathbb{R}} \phi^{-1} \left( \frac{2 \int_{-\infty}^{+\infty} \varphi_{\rho_0}(r) dr}{|a(t)|} \right) < +\infty.$$

So, $TB$ is uniformly bounded in $X$.

**Claim 2.2.** *$TB$ is equicontinuous on $X$.*

For $L > 0$, consider $t_1, t_2 \in [-L, L]$. Assume, without loss of generality, that $t_1 < t_2$.

Then, by (8.2.1), (8.2.5) and $(A_1)$,

$$|Tu(t_1) - Tu(t_2)| = \left| \int_{-\infty}^{t_1} \phi^{-1} \left( \frac{\tau_u + \int_{-\infty}^{s} f(r, u(r), u'(r)) dr}{a(s)} \right) ds \right.$$

$$\left. - \int_{-\infty}^{t_2} \phi^{-1} \left( \frac{\tau_u + \int_{-\infty}^{s} f(r, u(r), u'(r)) dr}{a(s)} \right) ds \right|$$

$$= \left| \int_{t_1}^{t_2} \phi^{-1} \left( \frac{\tau_u + \int_{-\infty}^{s} f(r, u(r), u'(r)) dr}{a(s)} \right) ds \right|$$

$$\leq \int_{t_1}^{t_2} \phi^{-1} \left( \frac{|\tau_u| + \int_{-\infty}^{s} |f(r, u(r), u'(r)) dr|}{|a(s)|} \right) ds$$

$$\leq \int_{t_1}^{t_2} \phi^{-1} \left( \frac{2 \int_{-\infty}^{+\infty} \varphi_{\rho_0}(r) dr}{|a(s)|} \right) ds$$

$$\longrightarrow 0, \quad \text{uniformly as } t_1 \to t_2,$$

and

$$|(Tu)'(t_1) - (Tu)'(t_2)| = \left| \phi^{-1} \left( \frac{\tau_u + \int_{-\infty}^{t_1} f(r, u(r), u'(r)) dr}{a(t_1)} \right) \right.$$

$$\left. - \phi^{-1} \left( \frac{\tau_u + \int_{-\infty}^{t_2} f(r, u(r), u'(r)) dr}{a(t_2)} \right) \right|$$

$$\longrightarrow 0, \quad \text{uniformly as } t_1 \to t_2.$$

Therefore, $TB$ is equicontinuous on $X$.

**Claim 2.3.** *$TB$ is equiconvergent at $\pm\infty$.*

Let $u \in B$. As in the previous claims

$$\left| Tu(t) - \lim_{t\to-\infty} (Tu(t)) \right| = \left| \int_{-\infty}^{t} \phi^{-1}\left( \frac{\tau_u + \int_{-\infty}^{s} f\left(r, u(r), u'(r)\right) dr}{a(s)} \right) ds \right|$$

$$\leq \int_{-\infty}^{t} \phi^{-1}\left( \frac{2\int_{-\infty}^{+\infty} \varphi_{\rho_0}(r) dr}{|a(s)|} \right) ds$$

$$\longrightarrow 0, \quad \text{as } t \to -\infty,$$

and

$$\left| Tu(t) - \lim_{t\to+\infty} (Tu(t)) \right| = \left| \int_{-\infty}^{t} \phi^{-1}\left( \frac{\tau_u + \int_{-\infty}^{s} f\left(r, u(r), u'(r)\right) dr}{a(s)} \right) ds \right.$$

$$\left. - \int_{-\infty}^{+\infty} \phi^{-1}\left( \frac{\tau_u + \int_{-\infty}^{s} f\left(r, u(r), u'(r)\right) dr}{a(s)} \right) ds \right|$$

$$= \left| \int_{t}^{+\infty} \phi^{-1}\left( \frac{\tau_u + \int_{-\infty}^{s} f\left(r, u(r), u'(r)\right) dr}{a(s)} \right) ds \right|$$

$$\leq \int_{t}^{+\infty} \phi^{-1}\left( \frac{2\int_{-\infty}^{+\infty} \varphi_{\rho_0}(r) dr}{|a(s)|} \right) ds$$

$$\longrightarrow 0, \quad \text{as } t \to +\infty.$$

Moreover,

$$\left| (Tu)'(t) - \lim_{t\to-\infty} (Tu)'(t) \right| = \left| \phi^{-1}\left( \frac{\tau_u + \int_{-\infty}^{t} f\left(r, u(r), u'(r)\right) dr}{a(t)} \right) \right.$$

$$\left. - \phi^{-1}\left( \frac{\tau_u}{\lim_{t\to-\infty} a(t)} \right) \right|$$

$$\leq \left| \phi^{-1}\left( \frac{\tau_u + \int_{-\infty}^{t} \varphi_{\rho_0}(r) dr}{a(t)} \right) \right|$$

$$\longrightarrow 0, \quad \text{as } t \to -\infty,$$

and

$$\left| (Tu)'(t) - \lim_{t \to +\infty} (Tu)'(t) \right| = \left| \phi^{-1} \left( \frac{\tau_u + \int_{-\infty}^{t} f(r, u(r), u'(r)) \, dr}{a(t)} \right) \right.$$

$$\left. - \phi^{-1} \left( \frac{\tau_u + \int_{-\infty}^{+\infty} f(r, u(r), u'(r)) \, dr}{\lim_{t \to +\infty} a(t)} \right) \right|$$

$$\longrightarrow 0, \quad \text{as } t \to +\infty.$$

So, $TB$ is equiconvergent at $\pm\infty$. By Lemma 7.2.2, $T$ is compact.

**Step 3.** Let $D \subset X$ be a closed bounded set. Then $TD \subset D$.

Suppose $D \subset X$ defined by

$$D = \{ x \in X : \|x\|_X \le \rho_1 \},$$

where $\rho_1$ is such that

$$\rho_1 := \max \left\{ \rho_0, \ |\nu^-| + \int_{-\infty}^{+\infty} \phi^{-1} \left( \frac{K}{a(s)} \right) ds, \sup_{t \in \mathbb{R}} \phi^{-1} \left( \frac{K}{a(t)} \right) \right\},$$

with

$$K := 2 \int_{-\infty}^{+\infty} \varphi_{\rho_0}(r) dr.$$

Let $u \in D$. By the same arguments as in Claim 2.1,

$$\|Tu\|_\infty = \sup_{t \in \mathbb{R}} |Tu(t)|$$

$$\le |\nu^-| + \int_{-\infty}^{+\infty} \phi^{-1} \left( \frac{|\tau_u| + \int_{-\infty}^{s} \varphi_{\rho_0}(r) dr}{|a(s)|} \right) ds$$

$$\le |\nu^-| + \int_{-\infty}^{+\infty} \phi^{-1} \left( \frac{2 \int_{-\infty}^{+\infty} \varphi_{\rho_0}(r) dr}{|a(s)|} \right) ds < \rho_1,$$

and

$$\|(Tu)'\|_\infty = \sup_{t \in \mathbb{R}} |(Tu)'(t)|$$

$$\le \sup_{t \in \mathbb{R}} \phi^{-1} \left( \frac{|\tau_u| + \int_{-\infty}^{t} |f(r, u(r), u'(r))| dr}{|a(t)|} \right)$$

$$\le \sup_{t \in \mathbb{R}} \phi^{-1} \left( \frac{2 \int_{-\infty}^{+\infty} \varphi_{\rho_0}(r) dr}{|a(t)|} \right) < \rho_1.$$

Therefore, $TD \subset D$. By Theorem 1.2.6, $T$ has at least one fixed point $u \in X$. That is, by Lemma 8.2.1, $u$ is a heteroclinic solution of (8.1.1).  $\square$

## 8.4.  Examples

**Example 1.** Consider the boundary value problem composed by the differential equation

$$[(t^2 + 1)\,(u'(t))^3]' = \frac{[(u(t))^2 - 1]\,e^{u'(t)}}{1 + t^2}, \quad \text{a.e. } t \in \mathbb{R}, \qquad (8.4.1)$$

and the boundary conditions

$$u(-\infty) = -1, \quad u(+\infty) = 1. \qquad (8.4.2)$$

This problem is a particular case of problem (8.1.1),(8.4.1), with

$$\phi(w) = w^3,$$

$$a(t) = 1 + t^2,$$

$$f(t, x, y) = \frac{(x^2 - 1)\,e^y}{1 + t^2},$$

$$\nu^- = -1, \quad \nu^+ = 1.$$

It can be seen that all assumptions of Theorem 8.3.1 are satisfied and $f$ is an $L^1$-Carathéodory function with

$$\varphi_\rho(t) = \frac{(\rho^2 + 1)\,e^\rho}{1 + t^2}.$$

Therefore, there is a heteroclinic connection linking the two equilibrium points $-1$ and $1$.

**Example 2.** The differential equation

$$[-(t^{2n} + 1)|u'(t)|^{p-2}u'(t)]' = \frac{[(u(t))^2 - 1][(u'(t))^6 + k]}{1 + t^4}, \quad \text{a.e. } t \in \mathbb{R},$$
$$(8.4.3)$$

where $n \in \mathbb{N}$, $k > 0$, and (8.4.2), is a particular case of problem (8.1.1),(8.1.2), with

$$\phi(y) = |y|^{p-2}y,$$

$$a(t) = -(1 + t^{2n}),$$

$$f(t, x, y) = \frac{(x^2 - 1)(y^6 + k)}{1 + t^4},$$

$$\nu^- = -1, \ \nu^+ = 1.$$

The assumptions of Theorem 9.2.6 are verified with

$$\varphi_\rho(t) = \frac{(\rho^2 + 1)(\rho^6 + 1)}{1 + t^4},$$

and there is a heteroclinic solution of (8.4.3) between the equilibrium points $-1$ and $1$.

## Chapter 9

# Heteroclinic Solutions for Semi-linear Problems (iii)

## 9.1. Introduction

In this chapter, we study the second-order nonautonomous half-linear equation on the whole real line,

$$(a\,(t, u(t))\,\phi\,(u'(t)))' = f\,(t, u(t), u'(t))\,, \quad \text{a.e. } t \in \mathbb{R}, \qquad (9.1.1)$$

with $\phi$ an increasing homeomorphism, $\phi(0) = 0$ and $\phi(\mathbb{R}) = \mathbb{R}$, $a \in C(\mathbb{R}^2, \mathbb{R})$ such that $a(t, x) > 0$ for $(t, x) \in \mathbb{R}^2$, and $f : \mathbb{R}^3 \to \mathbb{R}$ an $L^1$-Carathéodory function, together with the boundary conditions

$$u(-\infty) := \lim_{t \to -\infty} u(t) = \nu^-, \quad u(+\infty) := \lim_{t \to +\infty} u(t) = \nu^+, \qquad (9.1.2)$$

with $\nu^+, \nu^- \in \mathbb{R}$, such that $\nu^- < \nu^+$. Moreover, an application to singular $\phi$-Laplacian equations will be shown.

The problem (9.1.1),(9.1.2) was studied in [53, 112]. This chapter contains several results and criteria. For example, Theorem 2.1 guarantees the existence of a heteroclinic solution under, in short, the following main assumptions:

- $\phi$ grows at most linearly at infinity;
- $f(t, \nu^-, 0) \leq 0 \leq f(t, \nu^+, 0)$ for a.e. $t \in \mathbb{R}$;

- there exist constants $L, H > 0$, a continuous function $\theta : \mathbb{R}^+ \to \mathbb{R}^+$ and a function $\lambda \in L^p([-L, L])$, with $1 \le p \le \infty$, such that

$$|f(t, x, y)| \le \lambda(t)\, \theta\, (a(t, x)\, |y|)\,, \quad \text{for a.e. } |t| \le L, \text{ every } x \in \left[\nu^-, \nu^+\right],$$

$$|y| > H, \qquad \int^{+\infty} \frac{s^{1-\frac{1}{q}}}{\theta(s)} ds = +\infty;$$

- for every $C > 0$, there exist functions $\eta_C \in L^1(\mathbb{R})$, $\Lambda_C \in L^1_{\text{loc}}([0, +\infty))$, null in $[0, L]$ and positive in $[L, +\infty)$, and $N_C(t) \in L^1(\mathbb{R})$ such that

$$f(t, x, y) \le -\Lambda_C(t)\phi\,(|y|)\,,$$

$$f(-t, x, y) \ge \Lambda_C(t)\phi\,(|y|)\,, \quad \text{for a.e. } t \ge L, \text{ every } x \in \left[\nu^-, \nu^+\right],$$

$$|y| \le N_C(t)\,,$$

$$|f(t, x, y)| \le \eta_C(t) \text{ if } x \in \left[\nu^-, \nu^+\right], |y| \le N_C(t)\,, \quad \text{for a.e. } t \in \mathbb{R}.$$

Motivated by these works, we prove, in this paper, the existence of heteroclinic solutions for (9.1.1) assuming a Nagumo-type condition on the real line, and without asymptotic assumptions on the nonlinearities $\phi$ and $f$. The method follows arguments suggested in [71, 72, 115], applying the technique of [115] to a more general function $a$, to an adequate functional problem and to classical and singular $\phi$-Laplacian equations. The most common application for $\phi$ is the so-called $p$-Laplacian, that is $\phi(y) = |y|^{p-2}p$, $p > 1$, and even in this particular case verifies (9.1.3), the new assumption on $\phi$. On the other hand, to the best of our knowledge, the main result is even new when $\phi(y) = y$, that is, for equation

$$(a\,(t, u(t))\, u'(t))' = f(t, u(t), u'(t)), \quad \text{a.e. } t \in \mathbb{R}.$$

The study of differential equations and boundary value problems on the half-line or the whole real line and the existence of homoclinic or heteroclinic solutions have attracted increasing attention in the recent years due to the applications to non-Newtonian fluids theory, diffusion of flows in porous media, and nonlinear elasticity (see, for instance, [9, 25, 43, 85, 104, 110, 111, 135, 138, 140] and the references therein). In particular, heteroclinic connections are related to the processes in which the variable transits from an unstable equilibrium to a stable one (see, for example, [37, 42, 59, 61, 70, 108, 123, 127]), this is why heteroclinic solutions are often called transitional solutions.

Throughout this chapter, we consider the set $X := BC^1(\mathbb{R})$ of the $C^1(\mathbb{R})$ bounded functions, equipped with the norm $\|x\|_X = \max\{\|x\|_\infty, \|x'\|_\infty\}$, where $\|y\|_\infty := \sup_{t \in \mathbb{R}} |y(t)|$.

By using standard procedures, it can be shown that $(X, \|\cdot\|_X)$ is a Banach space.

As solution of problem (9.1.1),(9.1.2) we mean a function $u \in X$ such that $t \mapsto (a(t, u(t)) \phi(u'(t))) \in W^{1,1}(\mathbb{R})$, satisfying (9.1.1),(9.1.2).

The following hypotheses will be assumed:

$(H_1)$ $\phi$ is an increasing homeomorphism with $\phi(0) = 0$ and $\phi(\mathbb{R}) = \mathbb{R}$ such that

$$|\phi^{-1}(w)| \le \phi^{-1}(|w|); \tag{9.1.3}$$

$(H_2)$ $a \in C(\mathbb{R}^2, \mathbb{R})$ is a continuous and positive function with $a(t, x) \to +\infty$ as $|t| \to +\infty$.

## 9.2. Existence results

The first existence result for heteroclinic connections will be obtained for an auxiliary functional problem without the usual asymptotic or growth assumptions on $\phi$ or on the nonlinearity $f$.

Consider two continuous operators $A : X \to C(\mathbb{R})$, $x \longmapsto A_x$, with $A_x > 0$, $\forall x \in X$, and $F : X \to L^1(\mathbb{R})$, $x \longmapsto F_x$, and the functional problem composed by

$$(A_u(t) \, \phi(u'(t)))' = F_u(t), \quad \text{a.e. } t \in \mathbb{R}, \tag{9.2.1}$$

and the boundary conditions (9.1.2).

Define, for each bounded set $\Omega \subset X$,

$$m(t) := \min_{x \in \Omega} A_x(t) \tag{9.2.2}$$

and, for the above operators, assume that

$(F_1)$ For each $\eta > 0$ there is $\psi_\eta \in L^1(\mathbb{R})$, with $\psi_\eta(t) > 0$, a.e. $t \in \mathbb{R}$, such that $|F_x(t)| \le \psi_\eta(t)$, a.e. $t \in \mathbb{R}$, whenever $\|x\|_X < \eta$.

$(A_1)$ $A_x(t) \to +\infty$ as $|t| \to +\infty$ and

$$\int_{-\infty}^{+\infty} \phi^{-1}\left(\frac{2 \int_{-\infty}^{+\infty} \psi_\eta(r)dr}{m(s)}\right) ds < +\infty. \tag{9.2.3}$$

**Theorem 9.2.1.** *Assume that conditions $(H_1)$, $(F_1)$ and $(A_1)$ hold. Then there exists $u \in X$ such that $A_u \cdot (\phi \circ u') \in W^{1,1}(\mathbb{R})$ verifying (9.2.1) and (9.1.2).*

*Moreover, this solution is given by*

$$u(t) = \nu^- + \int_{-\infty}^{t} \phi^{-1} \left( \frac{\tau_u + \int_{-\infty}^{s} F_u(r)\,dr}{A_u(s)} \right) ds, \qquad (9.2.4)$$

*where $\tau_u$ is the unique solution of*

$$\int_{-\infty}^{+\infty} \phi^{-1} \left( \frac{\tau_u + \int_{-\infty}^{s} F_u(r)\,dr}{A_u(s)} \right) ds = \nu^+ - \nu^- \qquad (9.2.5)$$

*with*

$$\tau_u \in [w_1,\ w_3], \qquad (9.2.6)$$

*for*

$$w_1 := -\int_{-\infty}^{+\infty} |F_u(r)|\,dr, \qquad (9.2.7)$$

$$w_2 := \int_{-\infty}^{+\infty} |F_u(r)|\,dr, \ \ and\ w_3 = w_2 + k,\ (k \geq 0). \qquad (9.2.8)$$

**Proof.** For every $x \in X$, define the operator $T : X \to X$ by

$$T_x(t) = \nu^- + \int_{-\infty}^{t} \phi^{-1} \left( \frac{\tau_x + \int_{-\infty}^{s} F_x(r)\,dr}{A_x(s)} \right) ds,$$

where $\tau_x \in \mathbb{R}$ the unique solution of

$$\int_{-\infty}^{+\infty} \phi^{-1} \left( \frac{\tau_x + \int_{-\infty}^{s} F_x(r)\,dr}{A_x(s)} \right) ds = \nu^+ - \nu^-.$$

To show that $\tau_x$ is the unique solution of (9.2.5), consider the strictly increasing function in $\mathbb{R}$

$$G(y) := \int_{-\infty}^{+\infty} \phi^{-1} \left( \frac{y + \int_{-\infty}^{s} F_x(r)\,dr}{A_x(s)} \right) ds,$$

and remark that

$$\lim_{y \to -\infty} G(y) = \int_{-\infty}^{+\infty} \phi^{-1}(-\infty)\,ds = -\infty,$$

and

$$\lim_{y \to +\infty} G(y) = \int_{-\infty}^{+\infty} \phi^{-1}(+\infty)\, ds = +\infty. \tag{9.2.9}$$

Moreover, for $w_1$ given by (9.2.7) and $w_2$ given by (9.2.8), $G(w_1)$ and $G(w_2)$ have opposite signs, such as

$$G(w_1) = \int_{-\infty}^{+\infty} \phi^{-1}\left(\frac{w_1 + \int_{-\infty}^{s} F_x(r)\, dr}{A_x(s)}\right) ds \le 0 < \nu^+ - \nu^-,$$

$$G(w_2) = \int_{-\infty}^{+\infty} \phi^{-1}\left(\frac{w_2 + \int_{-\infty}^{s} F_x(r)\, dr}{A_x(s)}\right) ds \ge 0.$$

As $G$ is strictly increasing in $\mathbb{R}$, by (9.2.9), there is $k \ge 0$ such that $w_3 = w_2 + k$ and $G(w_3) \ge \nu^+ - \nu^-$. Therefore, the equation, $G(y) = \nu^- - \nu^+$, has a unique solution $\tau_x$, and by Bolzano's theorem, $\tau_x \in [w_1, w_3]$.

It is clear that if $T$ has a fixed point $u$, then $u$ is a solution of problem (9.2.1),(9.1.2).

To prove the existence of such fixed point, we consider several steps:

**Step 1.** $T : X \to X$ *is well defined*

with the positivity of $A$ and the continuity of $A$ and $F$, $T_x$ and

$$T_x'(t) = \phi^{-1}\left(\frac{\tau_x + \int_{-\infty}^{t} F_x(r)\, dr}{A_x(t)}\right)$$

are continuous on $\mathbb{R}$, that is, $T_x \in C^1(\mathbb{R})$.

Moreover, by $(H_1)$, $(F_1)$, $(A_1)$ and (9.2.5), $T_x$ and $T_x'$ are bounded. Therefore, $T_x \in X$.

**Step 2.** $T$ *is compact.*

Let $B \subset X$ be a bounded subset, $x \in B$, and $\rho_0 > 0$ such that $\|x\|_X < \rho_0$. Consider $m(t)$ given by (9.2.2) with $\Omega = B$.

**Claim.** $TB$ *is uniformly bounded in* $X$.

By (9.1.3), (9.2.6) and $(A_1)$, we have

$$\|T_x\|_\infty = \sup_{t \in \mathbb{R}} \left| \nu^- + \int_{-\infty}^{t} \phi^{-1}\left(\frac{\tau_x + \int_{-\infty}^{s} F_x(r)\, dr}{A_x(s)}\right) ds \right|$$

$$\le \sup_{t \in \mathbb{R}} \left( |\nu^-| + \int_{-\infty}^{t} \phi^{-1}\left( \left|\frac{\tau_x + \int_{-\infty}^{s} F_x(r)\, dr}{A_x(s)}\right| \right) ds \right)$$

$$\leq \sup_{t \in \mathbb{R}} \left( |\nu^-| + \int_{-\infty}^{t} \phi^{-1} \left( \frac{|\tau_x| + \int_{-\infty}^{s} |F_x(r)|}{A_x(s)} dr \right) ds \right)$$

$$\leq |\nu^-| + \int_{-\infty}^{+\infty} \phi^{-1} \left( \frac{|\tau_x| + \int_{-\infty}^{s} \psi_{\rho_0}(r) dr}{A_x(s)} \right) ds$$

$$\leq |\nu^-| + \int_{-\infty}^{+\infty} \phi^{-1} \left( \frac{2 \int_{-\infty}^{+\infty} \psi_{\rho_0}(r) dr + k}{m(s)} \right) ds < +\infty,$$

and

$$\|T_x'\|_\infty = \sup_{t \in \mathbb{R}} \left| \phi^{-1} \left( \frac{\tau_x + \int_{-\infty}^{t} F_x(r) \, dr}{A_x(t)} \right) \right|$$

$$\leq \sup_{t \in \mathbb{R}} \phi^{-1} \left( \frac{|\tau_x| + \int_{-\infty}^{t} |F_x(r)| \, dr}{A_x(t)} \right)$$

$$\leq \sup_{t \in \mathbb{R}} \phi^{-1} \left( \frac{|\tau_x| + \int_{-\infty}^{+\infty} \psi_{\rho_0}(r) dr}{A_x(t)} \right)$$

$$\leq \sup_{t \in \mathbb{R}} \phi^{-1} \left( \frac{2 \int_{-\infty}^{+\infty} \psi_{\rho_0}(r) dr + k}{m(t)} \right) < +\infty.$$

So, $TB$ is uniformly bounded in $X$.

**Claim.** *TB is equicontinuous on* $X$.

For $M > 0$, consider, $t_1, t_2 \in [-M, M]$, and, without loss of generality, $t_1 < t_2$.

Then, by (9.1.3), (9.2.6) and $(A_1)$,

$$|T_x(t_1) - T_x(t_2)| = \left| \int_{-\infty}^{t_1} \phi^{-1} \left( \frac{\tau_x + \int_{-\infty}^{s} F_x(r) \, dr}{A_x(s)} \right) ds \right.$$

$$- \int_{-\infty}^{t_2} \phi^{-1} \left( \frac{\tau_x + \int_{-\infty}^{s} F_x(r) \, dr}{A_x(s)} \right) ds \bigg|$$

$$= \left| \int_{t_1}^{t_2} \phi^{-1} \left( \frac{\tau_x + \int_{-\infty}^{s} F_x(r) \, dr}{A_x(s)} \right) ds \right|$$

$$\leq \int_{t_1}^{t_2} \phi^{-1} \left( \frac{|\tau_x| + \int_{-\infty}^{s} |F_x(r)| \, dr}{A_x(s)} \right) ds$$

$$\leq \int_{t_1}^{t_2} \phi^{-1} \left( \frac{2 \int_{-\infty}^{+\infty} \psi_{\rho_0}(r) dr + k}{m(s)} \right) ds$$

$$\longrightarrow 0, \text{ uniformly as } t_1 \to t_2,$$

and

$$|T_x'(t_1) - T_x'(t_2)| = \left| \phi^{-1} \left( \frac{\tau_x + \int_{-\infty}^{t_1} F_x(r) \, dr}{A_x(t_1)} \right) \right.$$

$$\left. - \phi^{-1} \left( \frac{\tau_x + \int_{-\infty}^{t_2} F_x(r) \, dr}{A_x(t_2)} \right) \right|$$

$$\longrightarrow 0, \text{ uniformly as } t_1 \to t_2.$$

Therefore, $TB$ is equicontinuous on $X$.

**Claim.** *$TB$ is equiconvergent at $\pm\infty$.*

Let $u \in B$. As in the claims above,

$$\left| T_x(t) - \lim_{t \to -\infty} (T_x(t)) \right| = \left| \int_{-\infty}^{t} \phi^{-1} \left( \frac{\tau_x + \int_{-\infty}^{s} F_x(r) \, dr}{A_x(s)} \right) ds \right|$$

$$\leq \int_{-\infty}^{t} \phi^{-1} \left( \frac{2 \int_{-\infty}^{+\infty} \psi_{\rho_0}(r) dr + k}{m(s)} \right) ds$$

$$\longrightarrow 0, \text{ as } t \to -\infty,$$

and

$$\left| T_x(t) - \lim_{t \to +\infty} (T_x(t)) \right| = \left| \int_{-\infty}^{t} \phi^{-1} \left( \frac{\tau_x + \int_{-\infty}^{s} F_x(r) \, dr}{A_x(s)} \right) ds \right.$$

$$\left. - \int_{-\infty}^{+\infty} \phi^{-1} \left( \frac{\tau_x + \int_{-\infty}^{s} F_x(r) \, dr}{A_x(s)} \right) ds \right|$$

$$= \left| \int_t^{+\infty} \phi^{-1} \left( \frac{\tau_x + \int_{-\infty}^s F_x(r)\,dr}{A_x(s)} \right) ds \right|$$

$$\leq \int_t^{+\infty} \phi^{-1} \left( \frac{2 \int_{-\infty}^{+\infty} \psi_\eta(r)\,dr + k}{m(s)} \right) ds$$

$$\longrightarrow 0, \text{ as } t \to +\infty.$$

Moreover, by $(A_1)$,

$$\left| T_x'(t) - \lim_{t \to -\infty} T_x'(t) \right| = \left| \phi^{-1} \left( \frac{\tau_x + \int_{-\infty}^t F_x(r)\,dr}{A_x(t)} \right) \right.$$

$$\left. -\phi^{-1} \left( \frac{\tau_x}{\lim_{t \to -\infty} A_x(t)} \right) \right|$$

$$\leq \left| \phi^{-1} \left( \frac{\tau_x + \int_{-\infty}^t \psi_{\rho_0}(r)\,dr}{A_x(t)} \right) \right|$$

$$\longrightarrow 0, \text{ as } t \to -\infty,$$

and

$$\left| T_x'(t) - \lim_{t \to +\infty} T_x'(t) \right| = \left| \phi^{-1} \left( \frac{\tau_x + \int_{-\infty}^t F_x(r)\,dr}{A_x(t)} \right) \right.$$

$$\left. - \phi^{-1} \left( \frac{\tau_x + \int_{-\infty}^{+\infty} F_x(r)\,dr}{\lim_{t \to -\infty} A_x(t)} \right) \right|$$

$$\longrightarrow 0, \text{ as } t \to +\infty.$$

So, $TB$ is equiconvergent at $\pm\infty$, and, by Lemma 7.2.2, $T$ is compact.

**Step 3.** *Let $D \subset X$ be a closed and bounded set. Then $TD \subset D$.*
Consider $D \subset X$ defined as

$$D = \{x \in X : \|x\|_X \leq \rho_1\},$$

with $\rho_1$, such that

$$\rho_1 := \max \left\{ \rho_0,\ |\nu^-| + \int_{-\infty}^{+\infty} \phi^{-1} \left( \frac{K}{m^*(s)} \right) ds, \sup_{t \in \mathbb{R}} \phi^{-1} \left( \frac{K}{m^*(t)} \right) \right\}$$

with

$$K := 2 \int_{-\infty}^{+\infty} \psi_{\rho_0}(r)dr + k,$$

and

$$m^*(t) := \min_{x \in B} A_x(t).$$

Let $x \in D$. Following similar arguments as in previous claims, with $m(t)$ given by (9.2.2) and $\Omega = D$,

$$\|T_x\|_\infty = \sup_{t \in \mathbb{R}} |T_x(t)|$$

$$\leq |\nu^-| + \int_{-\infty}^{+\infty} \phi^{-1} \left( \frac{|\tau_x| + \int_{-\infty}^{s} \psi_{\rho_0}(r)dr}{A_x(s)} \right) ds$$

$$\leq |\nu^-| + \int_{-\infty}^{+\infty} \phi^{-1} \left( \frac{2 \int_{-\infty}^{+\infty} \psi_{\rho_0}(r)dr + k}{m^*(s)} \right) ds < \rho_1,$$

and

$$\|T_x'\|_\infty = \sup_{t \in \mathbb{R}} |T_x'(t)| \leq \sup_{t \in \mathbb{R}} \phi^{-1} \left( \frac{|\tau_x| + \int_{-\infty}^{t} |F_x(r)|\, dr}{A_x(t)} \right)$$

$$\leq \sup_{t \in \mathbb{R}} \phi^{-1} \left( \frac{2 \int_{-\infty}^{+\infty} \psi_{\rho_0}(r)dr + k}{m^*(t)} \right) < \rho_1.$$

Therefore, $TD \subset D$. By Theorem 1.2.6, $T_x$ has a fixed point in $X$, that is, there is a heteroclinic solution of problem (9.2.1),(9.1.2). □

To make the relation between the functional problem and the initial one, we apply lower and upper solutions method, according to the following definition.

**Definition 9.2.2.** A function $\alpha \in X$ is a lower solution of problem (9.1.1),(9.1.2) if $t \mapsto (a(t, \alpha(t))\, \phi(\alpha'(t))) \in W^{1,1}(\mathbb{R})$,

$$(a(t, \alpha(t))\, \phi(\alpha'(t)))' \geq f(t, \alpha(t), \alpha'(t)), \quad \text{a.e. } t \in \mathbb{R}, \tag{9.2.10}$$

and

$$\alpha(-\infty) \leq \nu^-, \quad \alpha(+\infty) \leq \nu^+. \tag{9.2.11}$$

An upper solution $\beta \in X$ of problem (9.1.1),(9.1.2) satisfies $t \mapsto (a(t, \beta(t))$ $\phi(\beta'(t))) \in W^{1,1}(\mathbb{R})$ and the reversed inequalities.

To have some control on the first derivative, we apply a Nagumo-type condition.

**Definition 9.2.3.** An $L^1$-Carathéodory function $f : \mathbb{R}^3 \to \mathbb{R}$ satisfies a Nagumo-type growth condition relative to $\alpha, \beta \in X$, with $\alpha(t) \leq \beta(t)$, $\forall t \in \mathbb{R}$, if there are positive and continuous functions $\psi, \theta : \mathbb{R} \to \mathbb{R}^+$, such that

$$\sup_{t\in\mathbb{R}} \psi(t) < +\infty, \quad \int_0^{+\infty} \frac{|\phi^{-1}(s)|}{\theta(|\phi^{-1}(s)|)} ds = +\infty, \tag{9.2.12}$$

and

$$|f(t,x,y)| \leq \psi(t)\,\theta(|y|), \quad \text{for a.e. } t \in \mathbb{R} \quad \text{and} \quad \alpha(t) \leq x \leq \beta(t). \tag{9.2.13}$$

**Lemma 9.2.4.** *Let $f : \mathbb{R}^3 \to \mathbb{R}$ be an $L^1$-Carathéodory function $f : \mathbb{R}^3 \to \mathbb{R}$ satisfying a Nagumo-type growth condition relative to $\alpha, \beta \in BC(\mathbb{R})$, with $\alpha(t) \leq \beta(t)$, $\forall t \in \mathbb{R}$. Then there exists $N > 0$ (not depending on $u$) such that for every solution $u$ of (9.1.1), (9.1.2) with*

$$\alpha(t) \leq u(t) \leq \beta(t), \quad \text{for } t \in \mathbb{R}, \tag{9.2.14}$$

*we have*

$$\|u'\|_\infty < N. \tag{9.2.15}$$

**Proof.** Let $u$ be a solution of (9.1.1),(9.1.2) verifying (9.2.14). Take $r > 0$ such that

$$r > \max\left\{|\nu^-|, |\nu^+|\right\}. \tag{9.2.16}$$

If $|u'(t)| \leq r, \forall t \in \mathbb{R}$, the proof would be complete by taking $N > r$.
Suppose there is $t_0 \in \mathbb{R}$ such that $|u'(t_0)| > N$.
In the case $u'(t_0) > N$, by (9.2.12), we can take $N > r$ such that

$$\int_{a(t,u(t))\phi(r)}^{a(t,u(t))\phi(N)} \frac{|\phi^{-1}(\frac{s}{a(s,u(s))})|}{\theta(|\phi^{-1}(\frac{s}{a(s,u(s))})|)} ds > M\left(\sup_{t\in\mathbb{R}} \beta(t) - \inf_{t\in\mathbb{R}} \alpha(t)\right), \tag{9.2.17}$$

with $M := \sup_{t\in\mathbb{R}} \psi(t)$, which is finite by (9.2.12).

By (9.1.2), there are $t_1, t_2 \in \mathbb{R}$ such that $t_1 < t_2$, $u'(t_1) = N$, $u'(t_2) = r$ and $r \le u'(t) \le N, \forall t \in [t_1, t_2]$. So, the following contradiction with (9.2.17) holds by (9.2.12):

$$\int_{a(t,u(t))\phi(r)}^{a(t,u(t))\phi(N)} \frac{\left| \phi^{-1}\left( \frac{s}{a(s,u(s))} \right) \right|}{\theta\left( \left| \phi^{-1}\left( \frac{s}{a(s,u(s))} \right) \right| \right)} ds$$

$$= \int_{a(t,u(t))\phi(u'(t_2))}^{a(t,u(t))\phi(u'(t_1))} \frac{\left| \phi^{-1}\left( \frac{s}{a(s,u(s))} \right) \right|}{\theta\left( \left| \phi^{-1}\left( \frac{s}{a(s,u(s))} \right) \right| \right)} ds$$

$$= \int_{t_2}^{t_1} \frac{u'(s)}{\theta(u'(s))} \left( \phi\left( u'(s) \right) \right)' ds$$

$$= -\int_{t_1}^{t_2} \frac{f(s, u(s), u'(s))}{\theta(u'(s))} u'(s) ds$$

$$\le \int_{t_1}^{t_2} \frac{|f(s, u(s), u'(s))|}{\theta(u'(s))} u'(s) ds$$

$$\le \int_{t_1}^{t_2} \psi(s) u'(s) ds \le M \int_{t_1}^{t_2} u'(s) ds$$

$$\le M \left( u(t_2) - u(t_1) \right)$$

$$\le M \left( \sup_{t \in \mathbb{R}} \beta(t) - \inf_{t \in \mathbb{R}} \alpha(t) \right).$$

So, $u'(t) < N, \forall t \in \mathbb{R}$.

By similar arguments, it can be shown that $u'(t) > -N, \forall t \in \mathbb{R}$. Therefore, $\|u'\|_\infty < N, \forall t \in \mathbb{R}$. $\qquad \square$

The following lemma, in [141], provides a technical tool to use forward.

**Lemma 9.2.5.** *For $v, w \in C(I)$ such that $v(x) \le w(x)$, for every $x \in I$, define*

$$q(x, u) = \max\{v, \min\{u, w\}\}.$$

*Then, for each $u \in C^1(I)$, the next two properties hold:*

(a) $\frac{d}{dx}q(x, u(x))$ *exists for a.e. $x \in I$.*

(b) *If $u, u_m \in C^1(I)$ and $u_m \to u$ in $C^1(I)$, then*

$$\frac{d}{dx}q(x, u_m(x)) \to \frac{d}{dx}q(x, u(x)) \text{ for a.e. } x \in I.$$

The main result will be given by the following theorem.

**Theorem 9.2.6.** *Suppose that $f : \mathbb{R}^3 \to \mathbb{R}$ is an $L^1$-Carathéodory function verifying a Nagumo-type condition and hypothesis $(H_1)$, $(H_2)$. If there are lower and upper solutions of problem (9.1.1),(9.1.2), $\alpha$ and $\beta$, respectively, such that*

$$\alpha(t) \le \beta(t), \quad \forall t \in \mathbb{R},$$

*then there is a function $u \in X$ with $t \mapsto (a(t, u(t)) \phi(u'(t))) \in W^{1,1}(\mathbb{R})$, solution of problem (9.1.1),(9.1.2) and*

$$\alpha(t) \le u(t) \le \beta(t), \quad \forall t \in \mathbb{R}.$$

**Proof.** Define the truncation operator $Q : W^{1,1}(\mathbb{R}) \to X \subset W^{1,1}(\mathbb{R})$ given by

$$Q(x) := Q_x(t) = \begin{cases} \beta(t), & x(t) > \beta(t), \\ x(t), & \alpha(t) \le x(t) \le \beta(t), \\ \alpha(t), & x(t) < \alpha(t). \end{cases}$$

Consider the modified equation

$$\left( a(t, Q_u(t)) \, \phi\left( \frac{d}{dt}Q_u(t) \right) \right)' = f\left( t, Q_u(t), \frac{d}{dt}Q_u(t) \right)$$

$$+ \frac{1}{1+t^2} \frac{u(t) - Q_u(t)}{1 + |u(t) - Q_u(t)|}, \quad (9.2.18)$$

for a.e. $t \in \mathbb{R}$, which is well defined by Lemma 9.2.5.

**Claim 1.** *Every solution $u(t)$ of problem (9.2.18),(9.1.2) verifies*

$$\alpha(t) \le u(t) \le \beta(t), \quad \forall t \in \mathbb{R}.$$

Let $u$ be a solution of problem (9.2.18),(9.1.2), and suppose, by contradiction, that there is $t_0$ such that $\alpha(t_0) > u(t_0)$. Remark that, by (9.2.11), $t_0 \neq \pm\infty$ as $u(\pm\infty) - \alpha(\pm\infty) \geq 0$.

Define

$$\min_{t \in \mathbb{R}}(u(t) - \alpha(t)) := u(t_1) - \alpha(t_1) < 0.$$

So, there is an interval $]t_2, t_1]$ such that $u(t) - \alpha(t) < 0$, for a.e. $t \in ]t_2, t_1]$, and, by (9.2.10), this contradiction is achieved:

$$\left( a(t, \alpha(t)) \, \phi(\alpha'(t)) \right)' = \left( a(t, Q_u(t)) \, \phi\left( \frac{d}{dt} Q_u(t) \right) \right)'$$

$$= f\left( t, Q_u(t), \frac{d}{dt} Q_u(t) \right) + \frac{1}{1 + t^2} \frac{u(t) - Q_u(t)}{1 + |u(t) - Q_u(t)|}$$

$$< f(t, \alpha(t), \alpha'(t)) \leq \left( a(\alpha(t)) \, \phi(\alpha'(t)) \right)'.$$

Therefore, $\alpha(t) \leq u(t), \forall t \in \mathbb{R}$. Following similar arguments, it can be proved that $u(t) \leq \beta(t), \forall t \in \mathbb{R}$.

**Claim 2.** *Problem (9.2.18),(9.1.2) has a solution.*

Let $A : X \to C(\mathbb{R})$ and $F : X \to L^1(\mathbb{R})$ be the operators given by $A_x := a(t, Q_x(t))$ and

$$F_x := f\left( t, Q_x(t), \frac{d}{dt} Q_x(t) \right) + \frac{1}{1 + t^2} \frac{u(t) - Q_x(t)}{1 + |u(t) - Q_x(t)|}.$$

If, for

$$\rho := \max \left\{ \|\alpha\|_\infty, \|\beta\|_\infty, \|\alpha'\|_\infty, \|\beta'\|_\infty, N \right\},$$

with $N$ given by (9.2.15),

$$|F_x| \leq \left| f\left( t, Q_x(t), \frac{d}{dt} Q_x(t) \right) \right| + \frac{1}{1 + t^2} \frac{|u(t) - Q_x(t)|}{1 + |u(t) - Q_x(t)|}$$

$$\leq \left| f\left( t, Q_x(t), \frac{d}{dt} Q_x(t) \right) \right| \leq \varphi_\rho(t),$$

then $F_x$ verifies $(F_1)$. Moreover, from

$$a(t, Q_x(t)) \geq \min_{t \in \mathbb{R}} \left\{ a(t, \alpha(t)), a(t, \beta(t)) \right\},$$

we obtain that $A$ satisfies $(A_1)$ with $0 < m(t) \le \min_{t \in \mathbb{R}} \{a(t, \alpha(t)),$ $a(t, \beta(t))\}$.

So, by Theorem 9.2.1, problem (9.2.18),(9.1.2) has a solution, which, by Claim 1, is a solution of problem (9.1.1),(9.1.2).     □

## 9.3.   Example

Consider the boundary value problem, defined on the whole real line, composed by the differential equation

$$[((tu(t))^4 + 1)(u'(t))^3]' = \frac{[(u(t))^2 - 1](u'(t))^2}{1 + t^2}, \quad \text{a.e. } t \in \mathbb{R}, \quad (9.3.1)$$

coupled with the boundary conditions

$$u(-\infty) = -1, u(+\infty) = 1. \quad (9.3.2)$$

Remark that the null function is not a solution of problem (9.3.1),(9.3.2), which is a particular case of (9.1.1),(9.1.2), with

$$\phi(w) = w^3,$$
$$a(t, x) = 1 + (tx)^4,$$
$$f(t, x, y) = \frac{(x^2 - 1)y^2}{1 + t^2},$$
$$\nu^- = -1, \quad \text{and} \quad \nu^+ = 1.$$

All hypotheses of Theorem 9.2.6 are satisfied. In fact,

- $f$ is an $L^1$-Carathéodory function with

$$\varphi_\rho(t) = \frac{(\rho^2 + 1)\rho^2}{1 + t^2};$$

- $\phi(w)$ verifies $(H_1)$ and function $a(t, x)$ satisfies $(H_2)$;
- the constant functions $\alpha(t) \equiv -1$ and $\beta(t) \equiv k$, with $k \in [1, +\infty[$, are lower and upper solutions of problem (9.3.1),(9.3.2), respectively.
- $f(t, x, y)$ verifies a Nagumo-type condition for $-1 \le x \le k$ with

$$\psi(t) = \frac{k}{1 + t^2} \quad \text{and} \quad \theta(y) = y^2.$$

So, by Theorem 9.2.6, there is a heteroclinic connection $u$ between two equilibrium points $-1$ and $1$ of problem (9.3.1),(9.3.2), such that

$$-1 \le u(t) \le k, \quad \forall t \in \mathbb{R}, \quad k \ge 1.$$

## 9.4. Singular $\phi$-Laplacian equations

The previous theory can be easily adapted to singular $\phi$-Laplacian equations, such that for equations

$$\left(a\left(t, u(t)\right) \phi\left(u'(t)\right)\right)' = f\left(t, u(t), u'(t)\right), \quad \text{a.e. } t \in \mathbb{R}, \tag{1s}$$

where $\phi$ verifies

$(H_s)$  $\phi : (-b, b) \to \mathbb{R}$, for some $0 < b < +\infty$, this is an increasing homeomorphism with $\phi(0) = 0$ and $\phi(-b, b) = \mathbb{R}$ such that

$$\left|\phi^{-1}(w)\right| \le \phi^{-1}(|w|).$$

In this case, a heteroclinic solution of (1s), that is, a solution for problem (1s),(9.1.2), is a function $u \in X$ such that $u'(t) \in (-b, b)$, for $t \in \mathbb{R}$, and $t \mapsto \left(a\left(t, u(t)\right) \phi\left(u'(t)\right)\right) \in W^{1,1}(\mathbb{R})$, satisfying (1s),(9.1.2).

The theory for singular $\phi$-Laplacian equations is analogous to Theorems 9.2.1 and 9.2.6, replacing assumption $(H_1)$ with $(H_s)$.

As an example, we can consider the problem, for $n \in \mathbb{N}$ and $k > 0$,

$$\begin{cases} \left(\left((tu(t))^{2n} + 1\right) \dfrac{u'(t)}{\sqrt{1 - (u'(t))^2}}\right)' \\ \quad = \dfrac{\left((u(t))^2 - 1\right)\left(|u'(t)| + k\right)}{1 + t^2}, \quad \text{a.e. } t \in \mathbb{R}, \\ u(-\infty) = -1, \quad u(+\infty) = 1. \end{cases} \tag{9.4.1}$$

Clearly, problem (9.4.1) is a particularization of (9.1.1),(9.1.2), with

$$\phi(w) = \frac{w}{\sqrt{1 - w^2}}, \quad \text{for } w \in (-1, 1),$$

which models mechanical oscillations under relativistic effects,

$$a(t, x) = 1 + (tx)^{2n}, \tag{9.4.2}$$

$$f(t, x, y) = \frac{\left(x^2 - 1\right)\left(|y| + k\right)}{1 + t^2}, \tag{9.4.3}$$

$$\nu^- = -1, \quad \text{and} \quad \nu^+ = 1.$$

Moreover, the nonlinearity $f$ given by (9.4.3) is an $L^1$-Carathéodory function with

$$\varphi_\rho(t) = \frac{(\rho^2 + 1)(\rho + k)}{1 + t^2}.$$

Conditions of Theorem 9.2.6 are satisfied with $(H_1)$ replaced by $(H_s)$, such as

- the function $a(t, x)$, defined by (9.4.2), verifies $(H_2)$;
- the constant functions $\alpha(t) \equiv -1$ and $\beta(t) \equiv 1$ are lower and upper solutions of problem (9.4.1), respectively;
- $f(t, x, y)$ verifies a Nagumo-type condition for $-1 \leq x \leq 1$ with

$$\psi(t) = 1 \quad \text{and} \quad \theta(y) = |y| + k.$$

So, there is a heteroclinic connection $u$ between two equilibrium points $-1$ and $1$, for the singular $\phi$-Laplacian problem (9.4.1), such that

$$-1 \leq u(t) \leq 1, \quad \forall t \in \mathbb{R}.$$

Chapter 10

# Hammerstein Integral Equations with Sign-Changing Kernels

## 10.1. Introduction

In this chapter, we consider a Hammerstein generalized integral equation

$$u(t) = \int_{-\infty}^{+\infty} k(t,s) \, f(s, u(s), u'(s), \ldots, u^{(m)}(s)) \, ds, \qquad (10.1.1)$$

where $k : \mathbb{R}^2 \to \mathbb{R}$ is a $W^{m,\infty}(\mathbb{R}^2)$, $m \in \mathbb{N}$, kernel function and $f : \mathbb{R}^{m+2} \to \mathbb{R}$ is an $L^1$-Carathéodory function.

The existence of solutions of integral equations, in general, and Hammerstein equations, in particular, has been widely studied (see [18, 22, 41, 49, 84, 90, 148, 150], and the references therein). However, such equations where the nonlinearity can depend on the derivatives are scarce. In fact, this chapter considers discontinuous nonlinearities with derivative dependence, without monotone or asymptotic assumptions, on the whole real line.

We point out that the kernels, and their partial derivatives in order to the first variable, are very general functions: they may be discontinuous and may change sign. Moreover, the value of the limit of $k(t, s)$, when $|t| \to \infty$ provides an easy criterion to see if the existent solutions are homoclinic or heteroclinic.

The main tool to deal with the lack of compactness of the operator is the concept of equiconvergence at $\pm\infty$, suggested, for example, in [51, 128] (see Lemma 10.1.1). Our method for integral equations can be applied to boundary value problems which include differential equations of any order $n > m$.

In this sense, the last section of this chapter contains an application to a fourth-order nonlinear boundary value problem, which models moderately large deflections of infinite nonlinear beams resting on elastic foundations under localized external loads.

Along the chapter, $E := BC^m(\mathbb{R})$ is considered, the space of bounded and continuous functions on $\mathbb{R}$, with bounded and continuous derivatives on $\mathbb{R}$, till order $m$, equipped with the norm

$$\|u\|_E := \max\{\|u^{(j)}\|_\infty, j = 0, 1, \ldots, m\},$$

where $\|y\|_\infty := \sup_{t \in \mathbb{R}} |y(t)|$. For $W^{m,\infty}(\mathbb{R}^2)$, the space of functions in $L^\infty(\mathbb{R}^2)$, with derivatives, till order $m$, in $L^\infty(\mathbb{R}^2)$, we assume the following:

(A1) Function $k : \mathbb{R}^2 \to \mathbb{R}$ verifies $k \in W^{m,\infty}(\mathbb{R}^2)$,

$$\lim_{t \to \pm\infty} k(t,s) \in \mathbb{R}, \quad \lim_{t \to \pm\infty} \left| \frac{\partial^{(i)} k}{\partial t^i}(t,s) \right| \in \mathbb{R}, \quad \text{for } i = 1, \ldots, m, \ \forall s \in \mathbb{R},$$

and for all $\tau \in \mathbb{R}$,

$$\lim_{t \to \tau} \left| \frac{\partial^{(j)} k}{\partial t^j}(t,s) - \frac{\partial^{(j)} k}{\partial t^j}(\tau,s) \right| = 0, \quad \text{for a.e. } s \in \mathbb{R} \quad \text{and} \quad j = 0, \ldots, m.$$

(A2) There are positive functions $\psi_j : \mathbb{R} \to \mathbb{R}^+$ such that

$$\left| \frac{\partial^{(j)} k}{\partial t^j}(t,s) \right| \leq \psi_j(s) \quad \text{for } t \in \mathbb{R}, \text{ a.e. } s \in \mathbb{R} \quad \text{and } j = 0, \ldots, m$$

with

$$\int_{-\infty}^{+\infty} \psi_j(s) \varphi_r(s) ds < +\infty, \text{ for } j = 0, \ldots, m.$$

The following lemma (see [51, 128]) provides a compactness criterion to deal with the lack of compactness:

**Lemma 10.1.1.** *A set $M \subset X$ is relatively compact if the following conditions hold:*

(i) *$M$ is bounded in $X$:*

(ii) *the functions belonging to $M$ are equicontinuous on any compact interval of $\mathbb{R}$,*

(iii) *the functions from $M$ are equiconvergent at $\pm\infty$, that is, given $\epsilon > 0$, there exists $T(\epsilon) > 0$ such that*

$$|g^{(i)}(t) - g^{(i)}(+\infty)| < \epsilon \quad \text{and} \quad |g^{(i)}(t) - g^{(i)}(-\infty)| < \epsilon,$$

*for all $|t| > T(\epsilon), i = 0, 1, \ldots, m, \ m \in \mathbb{N}, \text{ and } g \in M.$*

## 10.2. Main result

The main theorem is as follows.

**Theorem 10.2.1.** *If* $f : \mathbb{R}^{m+2} \to \mathbb{R}$ *is an* $L^1$-*Carathéodory function and assumptions* (A1)–(A2) *hold, then problem* (10.1.1) *has at least one solution* $u(t) \in BC^m(\mathbb{R})$.

**Proof.** Define the continuous integral operator $T : E \to E$ given by

$$Tu(t) := \int_{-\infty}^{+\infty} k(t,s)\, f(s, u(s), u'(s), \dots, u^{(m)}(s))\, ds. \qquad (10.2.1)$$

Take $u \in E$. Then, there is $\rho > 0$ such that $\|u\|_E \leq \rho$.

To prove that the operator $T$ is compact, it is enough to show that the assumptions of Lemma 10.1.1 hold. For clarity, we divide the proof into several steps.

**Step 1.** $T$ *is well defined and uniformly bounded in* $E$.

As $f$ is $L^1$-Carathéodory, by the Lebesgue Dominated Convergence Theorem, (A1) and (A2), we have, for $i = 0, \dots, m$,

$$\| (Tu)^{(i)} \|_\infty = \sup_{t \in \mathbb{R}} \left| \int_{-\infty}^{+\infty} \frac{\partial^{(i)} k}{\partial t^i}(t,s) f(s, u(s), u'(s), \dots, u^{(m)}(s)) ds \right|$$

$$\leq \int_{-\infty}^{+\infty} \psi_i(s) \left| f(s, u(s), u'(s), \dots, u^{(m)}(s)) \right| ds$$

$$\leq \int_{-\infty}^{+\infty} \psi_i(s) \varphi_r(s) ds < +\infty.$$

Therefore $\|Tu\|_E < +\infty$, and, therefore, $TE \subset E$ and $T$ *is uniformly bounded in* $E$.

**Step 2.** $T$ *is equicontinuous in* $E$.

Consider $t_1, t_2 \in [0,1]$. By (A1),

$$|Tu(t_1) - Tu(t_2)|$$

$$\leq \int_{-\infty}^{+\infty} |k(t_1, s) - k(t_2, s)| \left| f(s, u(s), u'(s), \dots, u^{(m)}(s)) \right| ds$$

$$\leq \int_{-\infty}^{+\infty} |k(t_1, s) - k(t_2, s)| \varphi_r(s) ds \to 0, \text{ as } t_1 \to t_2,$$

and for $i = 1, \ldots, m$,

$$\left| (Tu)^{(i)}(t_1) - (Tu)^{(i)}(t_2) \right|$$

$$\leq \int_{-\infty}^{+\infty} \left| \frac{\partial^{(i)} k}{\partial t^i}(t_1, s) - \frac{\partial^{(i)} k}{\partial t^i}(t_2, s) \right| \left| f(s, u(s), u'(s), \ldots, u^{(n-1)}(s)) \right| ds$$

$$\leq \int_{-\infty}^{+\infty} \left| \frac{\partial^{(i)} k}{\partial t^i}(t_1, s) - \frac{\partial^{(i)} k}{\partial t^i}(t_2, s) \right| \varphi_r(s) ds \to 0, \quad \text{as } t_1 \to t_2.$$

Therefore, $T$ is equicontinuous in $E$.

**Step 3.** $T$ *is equiconvergent at* $\pm\infty$.

For $u \in E$, and for $i = 0, 1, \ldots, m$,

$$\left| (Tu(t))^{(i)} - \lim_{t \to \pm\infty} (Tu(t))^{(i)} \right|$$

$$= \left| \int_{-\infty}^{+\infty} \left( \frac{\partial^{(i)} k}{\partial t^i}(t, s) - \frac{\partial^{(i)} k}{\partial t^i}(\pm\infty, s) \right) f(s, u(s), u'(s), \ldots, u^{(m)}(s)) ds \right|$$

$$\leq \int_{-\infty}^{+\infty} \left| \frac{\partial^{(i)} k}{\partial t^i}(t, s) - \frac{\partial^{(i)} k}{\partial t^i}(\pm\infty, s) \right| \varphi_r(s) ds \to 0, \quad \text{as } t \to \pm\infty.$$

Then, by Lemma 10.1.1, $T$ is compact in $E$.

**Step 4.** $TD \subset D$, *for some* $D \subset X$ *a closed and bounded set.*

Consider a subset $D \subset X$ defined as

$$D := \left\{ u \in X : \|u\|_X \leq r_1 \right\},$$

with

$$r_1 := \max \left\{ r, \int_{-\infty}^{+\infty} \psi_i(s) \varphi_r(s) ds, \text{ for } i = 1, \ldots, m \right\},$$

where $r > 0$ is given by the $L^1$-bound of $f$.

Arguing as in Step 1, it can be shown that, for $i = 0, 1, \ldots, m$,

$$\| (Tu)^{(i)} \|_\infty \leq \int_{-\infty}^{+\infty} \psi_i(s) \varphi_r(s) ds \leq r_1.$$

Therefore, $TD \subset D$ and, by Schauder's fixed-point theorem, $T$ has at least a fixed point $u \in X$, which is the solution of equation (10.1.1). $\square$

**Corollary 10.2.2.** *If* $\lim_{t \to -\infty} k(t, s) = \lim_{t \to +\infty} k(t, s)$, *the solution of* (10.1.1) *is a homoclinic solution. If not, this solution is a heteroclinic solution.*

## 10.3.   Application to fourth-order BVPs and infinite beams

The integral equation (10.1.1) can be applied to boundary value problems of order $n$ such that $n > m$, defined on the whole real line.

As an application, we consider the case $n = 4$ and $m = 2$ through the study of infinite beams deflection.

Jany, in [87] considers the nonlinear Bernoulli–Euler–v. Karman problem composed by the fourth-order differential equation

$$EIu^{(4)}(t) + ku(t) = \frac{3}{2}EA(u'(t))^2 u''(t) + \omega(t), \quad t \in \mathbb{R}, \qquad (10.3.1)$$

and the boundary conditions

$$u(\pm\infty) := \lim_{t\to\pm\infty} u(t) = 0, \ u'(\pm\infty) := \lim_{t\to\pm\infty} u'(t) = 0. \qquad (10.3.2)$$

This problem is related to the analysis of moderately large deflections of infinite nonlinear beams resting on elastic foundations under localized external loads. More precisely, $E$ is the Young's modulus, $I$ the mass moment of inertia, $ku(t)$ the spring force upward, in which $k$ is a spring constant (for simplicity, the weight of the beam is neglected), $A$ the cross-sectional area of the beam and $\omega(t)$ the localized and applied loading downward.

As it was proved in [47], the above problem (10.3.1),(10.3.2) can be written as an integral equation

$$u(t) = \int_{-\infty}^{+\infty} G(t, s)f(s, u(s), u'(s), u''(s))ds, \qquad (10.3.3)$$

where $G(t, s)$ is the Green's function associated to (10.3.1),(10.3.2), defined by

$$G(t, s) = \frac{\sqrt[4]{\xi}}{2\xi} e^{-\frac{\sqrt[4]{\xi}|s-t|}{\sqrt{2}}} \sin\left(\frac{\sqrt[4]{\xi}|s-t|}{\sqrt{2}} + \frac{\pi}{4}\right), \qquad (10.3.4)$$

with $\xi = \frac{k}{EI}$.

By standard calculus, the following properties of the Green's function (10.3.4) can easily be obtained:

$$\lim_{|t|\to\infty} \frac{\partial^i G(t, s)}{\partial t^i} = 0, \ \left|\frac{\partial^i G(t, s)}{\partial t^i},\right| \leq \frac{\left(\sqrt[4]{\xi}\right)^{i+1}}{2\xi}, \quad \text{for } i = 0, 1, 2.$$

An example of this family of problems is given by the differential equation

$$u^{(4)}(t) + 3u(t) = \frac{3.4 + u^3(t) - u''(t)\,(u'(t))^2}{1 + t^4}, \quad t \in \mathbb{R}, \tag{10.3.5}$$

and the boundary conditions (10.3.2).

Here, the loading force $w(t) = \frac{3.4}{1+t^4}$ and the nonlinear function $g : \mathbb{R}^4 \to \mathbb{R}$ is defined by

$$g(t,x,y,z) := \frac{x^3 - zy^2}{1 + t^4}.$$

The function

$$f(t,x,y,z) := g(t,x,y,z) + w(t) \tag{10.3.6}$$

is $L^1$-Carathéodory and for $\max\{\|x\|, \|y\|, \|z\|\} < r, (r > 0)$, we have

$$\varphi_r(t) := \frac{3.4 + 2r^3}{1 + t^4}.$$

From the above, it is clear that (10.3.5) is a particular case of (10.1.1) for $k = 3$, $m = 2$, and the nonlinearity is given by (10.3.6). Moreover, the assumptions (A1) and (A2) are satisfied with

$$k(t,s) = G(t,s), \quad \psi_j(s) \equiv \frac{\left(\sqrt[4]{\xi}\right)^{i+1}}{2\xi}, \quad \text{for } j = 0,1,2.$$

Therefore, by Theorem 10.2.1, the integral equation (10.3.3) is a solution $u(t) \in \mathrm{BC}^2[0,1]$, which is a solution of the boundary value problem (10.3.5),(10.3.2). Moreover, from Corollary 10.2.2, as in (10.3.4)

$$\lim_{t \to -\infty} G(t,s) = \lim_{t \to +\infty} G(t,s) = 0,$$

this solution is a homoclinic solution of problem (10.3.5),(10.3.2).

# Part IV
# Functional Boundary Value Problems

# Introduction

Many phenomena of real life have a retrospective effect, i.e., their status in the future may depend not only on the present but also on what happened in the past. One of the mathematical processes appropriate to study this effect distributed over time is given by Functional Differential Equations (FDEs). It should be noted that the concept of FDEs generalizes the common differential equations into functions with a continuous argument.

Let us express the meaning of "functional" a little more. In Algebra, we deal with algebraic equations involving one or more unknown real numbers. Functional equations are much like algebraic equations, except that the unknown quantities are functions rather than real numbers.

From a historic point a view, as far as we know, the first time when functional equations were studied was in the fourteenth century in the work of mathematician Nicole Oresme (1323–1382) who provided an indirect definition of linear functions by means of a functional equation: in modern terminology, we have three distinct real numbers $x$, $y$, and $z$, and, associated to each one, a variable (the "intensity" of the quality at each point) which we can write as $f(x)$, $f(y)$, and $f(z)$, respectively (for more details, see [133]). The function $f$, considered as a linear function, is defined by the relation

$$\frac{y-x}{z-y} = \frac{f(y)-f(x)}{f(z)-f(y)}, \quad \text{for all distinct values of } x,y,z.$$

FDEs only appear, to the best of our knowledge, in the second half of the last century (see, for example, [58, 79, 91]).

However, the word "functional" was restricted to delay, advanced or neutral differential equations. This concept was adapted to a global unknown functional variable in, for instance, [31, 36]. If the functional part

135

appears in the differential equation, then it covers differential, integral or integro-differential equations, delay, neutral or advanced equations, among others. If the functional variation exists in the boundary conditions, then these boundary value problems include the classical two-point or multipoint conditions, and also nonlocal, integral boundary data, and cases where the global behavior of the unknown variable and its derivatives are involved. As an illustration of this type of functional problem with functional boundary conditions, we refer the problem in [114], with a functional variation in $u, u'$ and $u''$ in the differential equation,

$$-(\phi(u'''(x)))' = f(x, u''(x), u'''(x), u, u', u''),$$

for a.e. $x \in ]a, b[$, where $\phi$ is an increasing homeomorphism, $I := [a, b]$, and $f : I \times \mathbb{R}^2 \times (C(I))^3 \to \mathbb{R}^2$ is an $L^1$-Carathéodory function, and the boundary conditions

$$0 = L_1\left(u\left(a\right), u, u', u''\right),$$
$$0 = L_2\left(u'\left(a\right), u, u', u''\right),$$
$$0 = L_3\left(u''\left(a\right), u''\left(b\right), u'''\left(a\right), u'''\left(b\right), u, u', u''\right),$$
$$0 = L_4\left(u''\left(a\right), u''\left(b\right)\right),$$

where $L_i, i = 1, 2, 3, 4$, are suitable functions with $L_1$ and $L_2$ not necessarily continuous, satisfying some monotonicity assumptions.

In all the above references, functional boundary value problems are considered on bounded intervals. On unbounded domains, the techniques are more delicate due to the lack of compactness of the correspondent operators. For this reason, for example, the usual Arzèla–Ascoli theorem cannot be applied.

Part IV will present methods and techniques in order to consider some of these types of functional problems to unbounded domains, namely, the half-line or the whole real line.

In Chapter 11, an existence and localization result for a second-order BVP with functional boundary conditions will be proved. An application to an Emden–Fowler equation will be shown to illustrate the main result of the chapter.

Chapter 12 deals with third-order BVPs with functional boundary conditions. These types of problems can be observed, for example, in a Falkner–Skan equation and may describe the behavior of a viscous flow over a flat plate. The localization of a solution and, moreover, some of its qualitative properties will be presented in this chapter.

Chapter 13 covers the study of $\phi$-Laplacian equations. An existence and localization result will be proved and, in order to demonstrate the applicability of the main result, two examples will be shown.

# Chapter 11

# Second-Order Functional Problems

## 11.1. Introduction

Previous chapters have shown that some real phenomena are modeled by differential equations of various orders with different types of boundary conditions such as Sturm–Liouville, Homoclinic or Lidstone-type. There are, however, other problems with functional conditions, that is, situations where the boundary data do not depend on particular points but on the global variation of the unknown function. These may, for example, be provided with integral, differential, maximum or minimum arguments.

In order to cover a wide range of applications, in this chapter, we study the general second-order differential equation

$$u''(t) = f(t, u(t), u'(t)), \quad t \geq 0, \tag{11.1.1}$$

where $f : \mathbb{R}_0^+ \times \mathbb{R}^2 \to \mathbb{R}$ is a continuous function, coupled with the functional conditions

$$\begin{cases} L(u, u(0), u'(0)) = 0, \\ u'(+\infty) = B, \end{cases} \tag{11.1.2}$$

with $L : C(\mathbb{R}_0^+) \times \mathbb{R}^2 \to \mathbb{R}$ a continuous function, verifying some monotone assumption, $B \in \mathbb{R}$, and $u'(+\infty) := \lim_{t \to +\infty} u'(t)$.

Note that this functional dependence allows not only conditions on the boundary but also multipoint conditions, that is, requirements on one or more interior points.

BVP (11.1.1),(11.1.2) covers a huge variety of problems such as separated, multipoint, nonlocal, integrodifferential, periodic, anti-periodic with maximum or minimum arguments. For example, in the case of integral conditions, it covers problems that arise naturally in the description of physical phenomena, for instance, thermal conduction, semiconductor and hydrodynamic problems (see [29, 71, 88, 98, 121, 136, 146, 151, 153] and references therein).

In most cases, positive solutions are searched in compact intervals. However, results on the solvability of BVPs on unbounded intervals (half-line or real line) are scarce.

The main technique relies on the lower and upper solutions. Rather than the existence of bounded or unbounded solutions, their localization provides some qualitative data, like, for example, signal variation and behavior (see [33, 113]). Some results are concerned with the existence of bounded or positive solutions, as in [105, 147] and the references therein. For problem (11.1.1),(11.1.2), the existence of two types of solutions is proved, depending on $B$: if $B \neq 0$, the solution is unbounded; if $B = 0$, the solution is bounded.

This chapter is organized as, follows. First, some auxiliary results are defined such as the adequate space functions, some weighted norms, a criterion to overcome the lack of compactness, and the definition of lower and upper solutions. Section 11.3 contains the main result, an existence and localization theorem, whose proof combines lower and upper solutions technique with the fixed point theory. Finally, Sections 11.4 and 11.5 contain one example and an application to some problem composed by an Emden–Fowler-type equation with infinite multipoint conditions, which are not covered by the existent literature.

## 11.2.  Definitions and auxiliary results

Consider the space of admissible functions

$$X_F = \left\{ x \in C^1(\mathbb{R}_0^+) : \lim_{t \to +\infty} \frac{x(t)}{1+t} \in \mathbb{R}, \ \lim_{t \to +\infty} x'(t) \in \mathbb{R} \right\},$$

equipped with the norm $\|x\|_{X_F} = \max\{\|x\|_0, \|x'\|_1\}$, where

$$\|\omega\|_0 := \sup_{t \geq 0} \frac{|\omega(t)|}{1+t} \quad \text{and} \quad \|\omega'\|_1 := \sup_{t \geq 0} |\omega'(t)|.$$

In this way, $(X_F, \|\cdot\|_{X_F})$ is a Banach space.

Solutions of the linear problem associated to (11.1.1) and usual boundary conditions are defined with Green's function, which can be obtained by standard calculus.

**Lemma 11.2.1.** *Let* $th, h \in L^1(\mathbb{R}_0^+)$ *and* $A, B \in \mathbb{R}$. *Then the linear BVP*

$$\begin{cases} u''(t) = h(t), & t \geq 0, \\ u(0) = A, \\ u'(+\infty) = B \end{cases} \tag{11.2.1}$$

*has a unique solution in* $X_F$, *given by*

$$u(t) = A + Bt + \int_0^{+\infty} G(t,s)h(s)ds, \tag{11.2.2}$$

*where*

$$G(t,s) = \begin{cases} -s, & 0 \leq s \leq t, \\ -t, & t \leq s < +\infty. \end{cases} \tag{11.2.3}$$

**Proof.** If $u$ is a solution of problem (11.2.1), then the general solution for the differential equation is

$$u(t) = c_1 + c_2\, t + \int_0^t (t-s)h(s)ds,$$

where $c_1, c_2 \in \mathbb{R}$. Since $u$ should satisfy the boundary conditions, one has

$$c_1 = A, \quad c_2 = B - \int_0^{+\infty} h(s)ds.$$

The solution becomes

$$u(t) = A + Bt - t\int_0^{+\infty} h(s)ds + \int_0^t (t-s)h(s)ds,$$

and by computation,

$$u(t) = A + Bt + \int_0^{+\infty} G(t,s)h(s)ds$$

with $G$ given by (11.2.3).

Conversely, if $u$ is a solution of (11.2.2), it is easy to show that it satisfies the differential equation in (11.2.1). Also, $u(0) = A$ and $u'(+\infty) = B$. $\square$

The lack of compactness of $X_F$ is overcome by the following lemma which gives a general criterion for relative compactness, referred to in [3].

**Lemma 11.2.2.** *A set* $M \subset X_F$ *is relatively compact if the following conditions hold:*

(i)  *all functions from $M$ are uniformly bounded;*
(ii)  *all functions from $M$ are equicontinuous on any compact interval of $\mathbb{R}_0^+$;*
(iii)  *all functions from $M$ are equiconvergent at infinity, that is, for any given $\epsilon > 0$, there exists a $t_\epsilon > 0$ such that*

$$\left| \frac{x(t)}{1+t} - \lim_{t \to +\infty} \frac{x(t)}{1+t} \right| < \epsilon, \quad \left| x'(t) - \lim_{t \to +\infty} x'(t) \right| < \epsilon$$

*for all $t > t_\epsilon$ and $x \in M$.*

The functions considered as lower and upper solutions for the initial problem are defined as follows.

**Definition 11.2.3.** Given $B \in \mathbb{R}$, a function $\alpha \in X_F$ is a lower solution of problem (11.1.1),(11.1.2) if

$$\begin{cases} \alpha''(t) \geq f(t, \alpha(t), \alpha'(t)), & t \geq 0, \\ L(\alpha, \alpha(0), \alpha'(0)) \geq 0, \\ \alpha'(+\infty) < B. \end{cases}$$

A function $\beta \in X_F$ is an upper solution if it satisfies the reverse inequalities.

## 11.3.  Existence and localization results

In this section, the existence of at least one solution for the problem (11.1.1),(11.1.2) is proved, and, moreover, some localization data, following the arguments applied in [45] are given.

**Theorem 11.3.1.** *Let $f : \mathbb{R}_0^+ \times \mathbb{R}^2 \to \mathbb{R}$ be a continuous function, and for each $\rho > 0$, there exists a positive function $\varphi_\rho$ with $\varphi_\rho, t\varphi_\rho \in L^1(\mathbb{R}_0^+)$ such that for $(x(t), y(t)) \in \mathbb{R}^2$ with $\sup_{t \geq 0} \left\{ \frac{|x(t)|}{1+t}, |y(t)| \right\} < \rho$,*

$$|f(t, x, y)| \leq \phi_\rho(t), \quad t \geq 0. \tag{11.3.1}$$

*Moreover, if $L(x_1, x_2, x_3)$ is nondecreasing on $x_1$ and $x_3$ and there are $\alpha, \beta$, lower and upper solutions of (11.1.1),(11.1.2), respectively, such that*

$$\alpha(t) \leq \beta(t), \quad \forall t \geq 0, \tag{11.3.2}$$

*then problem* (11.1.1),(11.1.2) *has at least one solution* $u \in X_F$ *with* $\alpha(t) \leq u(t) \leq \beta(t)$ *for* $t \geq 0$.

**Proof.** Let $\alpha$ and $\beta$ be, respectively, lower and upper solutions of (11.1.1), (11.1.2) verifying (11.3.2). Consider the modified problem

$$
\begin{cases}
u''(t) = f(t, \delta(t, u(t)), u'(t)) + \dfrac{1}{1+t^3} \dfrac{u(t) - \delta(t, u(t))}{1 + |u(t) - \delta(t, u(t))|}, & t \geq 0, \\
u(0) = \delta(0, u(0) + L(u, u(0), u'(0))), \\
u'(+\infty) = B,
\end{cases}
$$

(11.3.3)

where $\delta : \mathbb{R}_0^+ \times \mathbb{R} \to \mathbb{R}$ is given by

$$
\delta(t, x) = \begin{cases}
\beta(t), & x > \beta(t), \\
x, & \alpha(t) \leq x \leq \beta(t), \\
\alpha(t), & x < \alpha(t).
\end{cases}
$$

For clarity, the proof will follow several steps.

**Step 1.** *If $u$ is a solution of* (11.3.3), *then $\alpha(t) \leq u(t) \leq \beta(t), \forall t \geq 0$.*

Let $u$ be a solution of the modified problem (11.3.3) and suppose, by contradiction, that there exists $t \geq 0$ such that $\alpha(t) > u(t)$. Therefore,

$$
\inf_{t \geq 0} (u(t) - \alpha(t)) < 0.
$$

If there is $t_* > 0$ such that

$$
\min_{t \geq 0} (u(t) - \alpha(t)) := u(t_*) - \alpha(t_*) < 0,
$$

one has $u'(t_*) = \alpha'(t_*)$ and $u''(t_*) - \alpha''(t_*) \geq 0$. By Definition 11.2.3, the following contradiction holds:

$$
0 \leq u''(t_*) - \alpha''(t_*)
$$

$$
= f(t_*, \delta(t_*, u(t_*)), u'(t_*)) + \frac{1}{1+t_*^3} \frac{u(t_*) - \delta(t_*, u(t_*))}{1 + |u(t_*) - \delta(t_*, u(t_*))|} - \alpha''(t_*)
$$

$$
= f(t_*, \alpha(t_*), \alpha'(t_*)) + \frac{1}{1+t_*^3} \frac{u(t_*) - \alpha(t_*)}{1 + |u(t_*) - \alpha(t_*)|} - \alpha''(t_*)
$$

$$
\leq \frac{u(t_*) - \alpha(t_*)}{1 + |u(t_*) - \alpha(t_*)|} < 0.
$$

So, $u(t) \geq \alpha(t), \forall t > 0$.

If the infimum is attained at $t = 0$, then

$$\min_{t \geq 0}(u(t) - \alpha(t)) := u(0) - \alpha(0) < 0.$$

As $u$ is solution of (11.3.3), by the definition of $\delta$, the following contradiction is achieved:

$$0 > u(0) - \alpha(0) = \delta(0, u(0) + L(u, u(0), u'(0))) - \alpha(0) \geq \alpha(0) - \alpha(0)$$

$$= 0.$$

If

$$\inf_{t \geq 0}(u(t) - \alpha(t)) := u(+\infty) - \alpha(+\infty) < 0,$$

then $u'(+\infty) - \alpha'(+\infty) \leq 0$. As $u$ is solution of (11.3.3), by Definition 11.2.3, the following contradiction holds:

$$0 \geq u'(+\infty) - \alpha'(+\infty) = B - \alpha'(+\infty) > 0.$$

Therefore, $u(t) \leq \alpha(t), \forall t \geq 0$.

In a similar way, it can be proved that $u(t) \geq \beta(t), \ \forall t \geq 0$.

**Step 2.** *Problem* (11.3.3) *has at least one solution.*

Let $u \in X_F$ and define the operator $T : X_F \to X_F$

$$Tu(t) = \Delta + Bt + \int_0^{+\infty} G(t, s)F_u(s)ds$$

with

$$F_u(s) := f(s, \delta(s, u(s)), u'(s)) + \frac{1}{1 + s^3} \frac{u(s) - \delta(s, u(s))}{1 + |u(s) - \delta(s, u(s))|},$$

$\Delta := \delta(0, u(0) + L(u, u(0), u'(0)))$ and $G$ is the Green function given by (11.2.3).

Therefore, problem (11.3.3) becomes

$$\begin{cases} u''(t) = F_u(t), t \geq 0, \\ u(0) = \Delta, \\ u'(+\infty) = B, \end{cases} \qquad (11.3.4)$$

and if $tF_u(t), F_u(t) \in L^1(\mathbb{R}_0^+)$, by Lemma 11.2.1, it is enough to prove that $T$ has a fixed point.

**Step 2.1.** *T is well defined.*

As $f$ is a continuous function, $Tu \in C^1(\mathbb{R}_0^+)$ and, by (11.3.1), for any $u \in X_F$ with $\rho > \max\{\|\alpha\|_{X_F}, \|\beta\|_{X_F}\}$,

$$\int_0^{+\infty} |F_u(s)|\, ds \leq \int_0^{+\infty} \left(\phi_\rho(s) + \frac{1}{1+s^3}\right) ds < +\infty,$$

that is, $F_u(t)$ and $tF_u(t) \in L^1(\mathbb{R}_0^+)$. By Lebesgue Dominated Convergence Theorem,

$$\lim_{t \to +\infty} \frac{(Tu)(t)}{1+t} = \lim_{t \to +\infty} \frac{\Delta + Bt}{1+t} + \int_0^{+\infty} \lim_{t \to +\infty} \frac{G(t,s)}{1+t} F_u(s)\, ds$$

$$\leq B + \int_0^{+\infty} \left(\phi_\rho(s) + \frac{1}{1+s^3}\right) ds < +\infty,$$

and analogously for

$$\lim_{t \to +\infty} (Tu)'(t) = B - \lim_{t \to +\infty} \int_t^{+\infty} F_u(s)\, ds = B < +\infty.$$

Therefore, $Tu \in X_F$.

**Step 2.2.** *T is continuous.*

Consider a convergent sequence $u_n \to u$ in $X_F$, there exists $\rho_1 > 0$ such that $\max\{\|\alpha\|_{X_F}, \|\beta\|_{X_F}\} < \rho_1$.

With $M := \sup_{t \geq 0} \frac{|G(t,s)|}{1+t}$, one has

$$\|Tu_n - Tu|_{X_F} = \max\{\|Tu_n - Tu|_0, \|(Tu_n)' - (Tu)'\|_1\}$$

$$\leq \int_0^{+\infty} M|F_{u_n}(s) - F_u(s)|\, ds$$

$$+ \int_t^{+\infty} |F_{u_n}(s) - F_u(s)|\, ds \longrightarrow 0,$$

as $n \to +\infty$.

**Step 2.3.** *T is compact.*

Let $B \subset X_F$ be any bounded subset. Therefore, there is $r > 0$ such that $\|u\|_{X_F} < r$, $\forall u \in B$.

For each $u \in B$, and for $\max\{r, \|\alpha\|_{X_F}, \|\beta\|_{X_F}\} < r_1$

$$\|Tu\|_0 = \sup_{t \geq 0} \frac{|Tu(t)|}{1+t} \leq \sup_{t \geq 0} \frac{|\Delta + Bt|}{1+t}$$

$$+ \int_0^{+\infty} \sup_{t \geq 0} \frac{|G(t,s)|}{1+t} |F_u(s)| ds$$

$$\leq \sup_{t \geq 0} \frac{|\Delta + Bt|}{1+t} + \int_0^{+\infty} M\left(\phi_{r_1}(s) + \frac{1}{1+s^3}\right) ds < +\infty,$$

$$\|(Tu)'\|_1 = \sup_{t \geq 0} |(Tu)'(t)| \leq |B| + \int_t^{+\infty} |F_u(s)| ds$$

$$\leq |B| + \int_t^{+\infty} \left(\phi_{r_1}(s) + \frac{1}{1+s^3}\right) ds < +\infty.$$

So, $\|Tu\|_{X_F} = \max\{\|Tu\|_0, \|(Tu)'\|_1\} < +\infty$, that is, $TB$ is uniformly bounded in $X_F$.

$TB$ is equicontinuous because, for $L > 0$ and $t_1, t_2 \in (0, L]$, one has, as $t_1 \to t_2$,

$$\left| \frac{Tu(t_1)}{1+t_1} - \frac{Tu(t_2)}{1+t_2} \right|$$

$$\leq \left| \frac{\Delta + Bt_1}{1+t_1} - \frac{\Delta + Bt_2}{1+t_2} \right|$$

$$+ \int_0^{+\infty} \left| \frac{G(t_1,s)}{1+t_1} - \frac{G(t_2,s)}{1+t_2} \right| |F(u(s))| ds$$

$$\leq \left| \frac{\Delta + Bt_1}{1+t_1} - \frac{\Delta + Bt_2}{1+t_2} \right|$$

$$+ \int_0^{+\infty} \left| \frac{G(t_1,s)}{1+t_1} - \frac{G(t_2,s)}{1+t_2} \right| \left(\phi_{r_1}(s) + \frac{1}{1+s^3}\right) ds \longrightarrow 0,$$

$$|(Tu)'(t_1) - (Tu)'(t_2)| = \left| \int_{t_1}^{+\infty} F_u(s) ds - \int_{t_2}^{+\infty} F_u(s) ds \right|$$

$$\leq \int_{t_1}^{t_2} |F_u(s)| ds \leq \int_{t_1}^{t_2} \left(\phi_{r_1}(s) + \frac{1}{1+s^3}\right) ds \longrightarrow 0.$$

So, $TB$ is equicontinuous.

Moreover, $TB$ is equiconvergent at infinity because, as $t \to +\infty$,

$$\left| \frac{Tu(t)}{1+t} - \lim_{t \to +\infty} \frac{Tu(t)}{1+t} \right|$$

$$\leq \left| \frac{\Delta + Bt}{1+t} - B \right| + \int_0^{+\infty} \left| \frac{G(t,s)}{1+t} + 1 \right| |F_u(s)| ds$$

$$\leq \left| \frac{\Delta + Bt}{1+t} - B \right| + \int_0^{+\infty} \left| \frac{G(t,s)}{1+t} + 1 \right| \left( \phi_{\rho_1} + \frac{1}{1+s^3} \right) ds \to 0,$$

and

$$\left| (Tu)'(t) - \lim_{t \to +\infty} (Tu)'(t) \right| \leq \int_t^{+\infty} |F_u(s)| ds$$

$$\leq \int_t^{+\infty} \left( \rho_1 + \left( \frac{1}{(1+s^3)} \right) \right) ds \longrightarrow 0, \text{ as } t \to +\infty.$$

So, by Lemma 11.2.2, $TB$ is relatively compact.

Then by Schauder's fixed-point theorem (Theorem 1.2.6), $T$ has at least one fixed point $u_1 \in X_F$.

**Step 3.** *$u_1$ is a solution of problem* (11.1.1),(11.1.2).

By Step 1, if $u_1$ is a solution of (11.3.3), then $\alpha(t) \leq u_1(t) \leq \beta(t)$ for all $t \geq 0$. So, the differential equation (11.1.1) is obtained. It remains to be proved that $\alpha(0) \leq u_1(0) + L(u_1, u_1(0), u_1'(0)) \leq \beta(0)$.

Suppose, by contradiction, that $\alpha(0) > u_1(0) + L(u_1, u_1(0), u_1'(0))$. Then

$$u_1(0) = \delta(0, u_1(0) + L(u_1, u_1(0), u_1'(0))) = \alpha(0)$$

and by the monotony of $L$ and Definition 11.2.3, the following contradiction holds:

$$0 > u_1(0) + L(u_1, u_1(0), u_1'(0)) - \alpha(0)$$

$$= L(u_1, \alpha(0), u_1'(0)) \geq L(\alpha, \alpha(0), \alpha'(0)) \geq 0.$$

So, $\alpha(0) \leq u_1(0) + L(u_1, u_1(0), u_1'(0))$ and in a similar way, it can be proved that $u_1(0) + L(u_1, u_1(0), u_1'(0)) \leq \beta(0)$.

Therefore, $u_1$ is a solution of (11.1.1), (11.1.2). $\qquad \square$

A similar result can be obtained if $f$ is an $L^1$-Carathéodory function and equation (11.1.1) is replaced by

$$u''(t) = f(t, u(t), u'(t)), \quad \text{a.e. } t \geq 0. \tag{11.3.5}$$

However, in this case, an extra assumption on $f$ must be assumed.

**Theorem 11.3.2.** *Let $f : \mathbb{R}_0^+ \times \mathbb{R}^2 \to \mathbb{R}$ be an $L^1$-Carathéodory function such that $f(t, x, y)$ is monotone on $y$. If there are $\alpha, \beta$, lower and upper solutions of (11.3.5),(11.1.2), respectively, verifying (11.3.2) and $L(x_1, x_2, x_3)$ is nondecreasing on $x_1$ and $x_3$, then problem (11.3.5),(11.1.2) has at least one solution $u \in X_F$ with $\alpha(t) \leq u(t) \leq \beta(t)$, $\forall t \geq 0$.*

**Proof.** The proof is similar to Theorem 11.3.1 except the first step.

Let $u$ be a solution of the modified problem composed by

$$u''(t) = f(t, \delta(t, u(t)), u'(t)) + \frac{1}{1 + t^3} \frac{u(t) - \delta(t, u(t))}{1 + |u(t) - \delta(t, u(t))|}, \quad \text{a.e. } t \geq 0,$$

and the boundary conditions

$$u(0) = \delta(0, u(0) + L(u, u(0), u'(0))),$$

$$u'(+\infty) = B.$$

If, by contradiction, there is $t_* > 0$ such that

$$\min_{t \geq 0}(u(t) - \alpha(t)) := u(t_*) - \alpha(t_*) < 0,$$

then $u'(t_*) = \alpha'(t_*), u''(t_*) - \alpha''(t_*) \geq 0$, and there exists an interval $I_- := ]t_-, t_*[$ where $u(t) < \alpha(t)$, $u'(t) \leq \alpha'(t)$, $\forall t \in I_-$.

By Definition 11.2.3 and if $f(t, x, y)$ is nondecreasing on $y$, this contradiction holds for $t \in I_-$:

$$0 \leq u''(t) - \alpha''(t)$$

$$= f(t, \delta(t, u(t)), u'(t)) + \frac{1}{1 + t^3} \frac{u(t) - \delta(t, u(t))}{1 + |u(t) - \delta(t, u(t))|} - \alpha''(t)$$

$$\leq f(t, \alpha(t), \alpha'(t)) + \frac{1}{1 + t^3} \frac{u(t) - \alpha(t)}{1 + |u(t) - \alpha(t)|} - \alpha''(t)$$

$$\leq \frac{u(t) - \alpha(t)}{1 + |u(t) - \alpha(t)|} < 0.$$

The same remains valid if $f$ is nonincreasing, considering an interval $I_+ :=]t_*, t_+[$ where $u(t) < \alpha(t)$, $u'(t) \geq \alpha'(t)$, $\forall t \in I_+$.

So, in both cases, $u(t) \geq \alpha(t), \forall t \geq 0$.

The remaining steps are identical to the proof of Theorem 11.3.1, and it will be omitted. $\qquad\qquad\qquad\qquad\qquad\qquad\qquad\qquad\qquad\qquad\qquad\quad$ $\square$

## 11.4. Example

Consider the second-order problem on the half-line with functional boundary conditions

$$
\begin{cases}
u''(t) = \dfrac{\sin(u(t)+1) + (u'(t))^3 + u(t)e^{-t}}{1+t^3}, & t \geq 0, \\
4u^2(0) + \min\limits_{t\geq 0} u(t) + u'(0) - 2 = 0, & (11.4.1) \\
u'(+\infty) = 0,5.
\end{cases}
$$

Remark that the above problem is a particular case of (11.1.1), (11.1.2) with

$$
f(t,x,y) = \frac{\sin(x+1) + y^3 + xe^{-t}}{1+t^3},
$$

$$
B = 0,5,
$$

$$
L(a,b,c) = 4b^2 + \min_{t\geq 0} a(t) + c - 2.
$$

If $f$ is continuous in $\mathbb{R}_0^+$, then for $u \in X_F$, assumption (11.3.1) holds, with $\varphi_\rho = \frac{k}{1+t^3}$, for some $k > 0$ and $\rho > 1$.

The function $L(a,b,c)$ is not decreasing in $a$ and $c$, and $\alpha(t) \equiv -1$ and $\beta(t) = t$ are lower and upper solutions for (11.4.1), respectively, then, by Theorem 11.3.1, there is at least an unbounded solution $u$ of (11.4.1) such that

$$
-1 \leq u(t) \leq t, \quad \forall t \geq 0.
$$

## 11.5. Emden–Fowler equation

Emden–Fowler-type equations (see [144]) can model the heat diffusion perpendicular to parallel planes by

$$
\frac{\partial^2 u(x,t)}{\partial x^2} + \frac{\alpha}{x}\frac{\partial u(x,t)}{\partial x} + af(x,t)g(u) + h(x,t) = \frac{\partial u(x,t)}{\partial t}, \quad 0 < x < t,
$$

where $f(x,t)g(u) + h(x,t)$ means the nonlinear heat source and $u(x,t)$ gives the temperature at time $t$.

In the steady-state case, and with $h(x,t) \equiv 0$, last equation becomes

$$u''(x) + \frac{\alpha}{x}u'(x) + af(x)g(u) = 0, \quad x \geq 0. \tag{11.5.1}$$

If $f(x) \equiv 1$ and $g(u) = u^n$, then (11.5.1) is called the Lane–Emden equation of the first kind, whereas in the second kind, one has $g(u) = e^u$. Both cases are used in the study of thermal explosions. For more details, see [82].

In the literature, Emden–Fowler-type equations are associated to Dirichlet or Neumann boundary conditions (see [78, 142]). To the best of author's knowledge, this is the first time when some Emden–Fowler-type equations are considered together with functional boundary conditions on the half-line.

Consider that one looks for nonnegative solutions for the problem composed by the discontinuous differential equation

$$u''(x) = \frac{u'(x)}{1+x^3} + \frac{u^4(x)}{e^x}, \quad \text{a.e. } x > 0, \tag{11.5.2}$$

coupled with the infinite multipoint conditions

$$\begin{cases} \sum_{n=1}^{+\infty} a_n u(\eta_n) - u(0) + u'(0) = 0, \\ u'(+\infty) = \delta \quad (0 < \delta < 1), \end{cases} \tag{11.5.3}$$

where $a_n$ and $\eta_n$ are nonnegative sequences such that

$$a_1\eta_1 \geq a_2\eta_2 \geq \cdots \geq a_n\eta_n \geq \cdots, \quad \sum_{n=1}^{+\infty} a_n u(\eta_n) \quad \text{and} \quad \sum_{n=1}^{+\infty} a_n\eta_n$$

are convergent with $\sum_{n=1}^{+\infty} a_n(\eta_n + k) \leq 1 - k$ $(0 < k < 1)$.

This is a particular case of (11.3.5), (11.1.2), where

$$f(x,y,z) = \frac{z}{1+x^3} + \frac{y^4}{e^x},$$

$$B = \delta,$$

$$L(v,y,z) = \sum_{n=1}^{+\infty} a_n v(\eta_n) - y + z.$$

$$|f(x,y,z)| \leq \frac{k_1}{1+x^3} + \frac{k_2}{e^x} := \varphi_r(x), \; k_1, k_2 > 0, \; r > 1.$$

If $\varphi_r(x), x\varphi_r(x) \in L^1(\mathbb{R}_0^+)$, then $f$ is $L^1$-Carathéodory, and, moreover, $f$ is monotone on $z$ (is nondecreasing).

If $L(v, y, z)$ is not decreasing in $v$ and $z$, and functions $\alpha(x) \equiv 0$ and $\beta(x) = x + k$ are lower and upper solutions of problem (11.5.2),(11.5.3), respectively, then, by Theorem 11.3.2, there is at least an unbounded and nonnegative solution $u$ of (11.5.2),(11.5.3) such that

$$0 \le u(x) \le x + k, \quad \forall x \ge 0.$$

# Chapter 12

# Third-Order Functional Problems

## 12.1. Introduction

In this chapter, we consider a third-order BVP, composed by a fully differential equation

$$u'''(t) = f(t, u(t), u'(t), u''(t)), \quad t \geq 0, \qquad (12.1.1)$$

where $f : \mathbb{R}_0^+ \times \mathbb{R}^3 \to \mathbb{R}$ is an $L^1$-Carathéodory function, and the functional boundary conditions on the half-line

$$L_0(u, u(0)) = 0,$$

$$L_1(u, u'(0)) = 0, \qquad (12.1.2)$$

$$L_2(u, u''(+\infty)) = 0,$$

with $L_i : C(\mathbb{R}_0^+) \times \mathbb{R} \to \mathbb{R}, i = 0, 1, 2$ continuous functions, verifying some monotone assumptions and

$$u''(+\infty) := \lim_{t \to +\infty} u''(t).$$

There is an extensive literature on BVP defined in bounded domains as this type of problem is an adequate tool to describe countless phenomena of real life, such as models on chemical engineering, heat conduction, thermoelasticity, plasma physics, fluids flow, etc. (see, for instance, [30, 60, 69, 81, 88, 93, 106, 124]). However, on the real line or half-line, the results are very scarce (see, for example, [3, 151] and the references therein).

In some backgrounds, the models require different kinds of nonlocal or integral boundary conditions. In this way, it is useful to consider generalized boundary data, which include usual and nonclassic boundary conditions. In fact, if BVP contains a functional dependence on the unknown functions, or in its derivatives, either in the differential equation, or in the boundary data, these functional BVPs allow many more varieties of problems such as separated, multipoint, nonlocal, integro-differential, with maximum or minimum arguments, etc., as it can be seen, for instance, in [35, 38, 65, 71, 72, 121].

To the author's best knowledge, it is the first time where these types of functional boundary conditions are applied to third-order BVP on the half-line. From the different arguments used, weighted norms, fixed point theory and lower and upper solutions method can be highlighted. This last technique provides a location result, which is particularly useful to get some qualitative properties on the solution, such as positivity, monotony, convexity, etc.

The chapter is organized as follows: in the first section, some auxiliary results are defined such as the adequate space of admissible functions, the weighted norms, an existence result for a linear BVP via Green's functions, an *a priori* bound for the second derivative from a Nagumo-type condition, a criterion to overcome the lack of compactness, and the definition of lower and upper solutions. Section 12.3 contains the main result of the chapter — an existence and localization theorem, whose proof combines lower and upper solutions technique with the fixed point theory. Finally, an application to a Falkner–Skan equation is shown to illustrate the main result, which is not covered by previous works in the literature as far as we know.

## 12.2. Definitions and *a priori* bounds

Consider the space of admissible functions

$$
X_{F3} = \left\{
\begin{array}{l}
x \in C^2(\mathbb{R}_0^+) : \lim\limits_{t \to +\infty} \dfrac{x(t)}{1+t^2} \in \mathbb{R}, \\[2mm]
\lim\limits_{t \to +\infty} \dfrac{x'(t)}{1+t} \in \mathbb{R}, \ \lim\limits_{t \to +\infty} x''(t) \in \mathbb{R}
\end{array}
\right\},
$$

with the norm $\|x\|_{X_{F3}} = \max\{\|x\|_0, \|x'\|_1, \|x''\|_2\}$, where

$$
\|\omega\|_0 := \sup_{t \ge 0} \frac{|\omega(t)|}{1+t^2}, \ \|\omega\|_1 := \sup_{t \ge 0} \frac{|\omega(t)|}{1+t} \ \text{and} \ \|\omega\|_2 := \sup_{t \ge 0} |\omega(t)|.
$$

Defining in this way, $(X_{F3}, \|\cdot\|_{X_{F3}})$ is a Banach space.

The solutions of the linear problem associated to (12.1.1), with the two-point boundary conditions on the half line, can be defined with Green's function.

**Lemma 12.2.1.** *Let* $t^2h, th, h \in L^1(\mathbb{R}_0^+)$. *Then the linear BVP*

$$\begin{cases} u'''(t) = h(t), & \text{a.e. } t \geq 0, \\ u(0) = A, \\ u'(0) = B, \\ u''(+\infty) = C, \end{cases} \tag{12.2.1}$$

*with* $A, B, C \in \mathbb{R}$, *has a unique solution given by*

$$u(t) = A + Bt + \frac{Ct^2}{2} + \int_0^{+\infty} G(t,s)h(s)ds, \tag{12.2.2}$$

*where*

$$G(t,s) = \begin{cases} \dfrac{s^2}{2} - ts, & 0 \leq s \leq t, \\ -\dfrac{t^2}{2}, & 0 \leq t \leq s < +\infty. \end{cases} \tag{12.2.3}$$

**Proof.** If $u$ is a solution of problem (12.2.1), then the general solution for the differential equation is

$$u(t) = c_1 + c_2 t + c_3 t^2 + \int_0^t \left( \frac{s^2}{2} - ts + \frac{t^2}{2} \right) h(s)ds,$$

where $c_1, c_2, c_3$ are real constants. Since $u(t)$ should satisfy the boundary conditions,

$$c_1 = A, \quad c_2 = B, \quad c_3 = \frac{C}{2} - \frac{1}{2} \int_0^{+\infty} h(s)ds,$$

and, therefore,

$$u(t) = A + Bt + \frac{Ct^2}{2} - \frac{t^2}{2} \int_0^{+\infty} h(s)ds + \int_0^t \left( \frac{s^2}{2} - ts + \frac{t^2}{2} \right) h(s)ds,$$

which can be written as (12.2.2) with $G(t,s)$ given by (12.2.3). □

Some trivial properties of (12.2.3) will play an important role forward.

**Lemma 12.2.2.** *Function* $G(t,s)$ *defined by* (12.2.3) *verifies*

(i) $\displaystyle \lim_{t \to +\infty} \frac{G(t,s)}{1+t^2} \in \mathbb{R}, \quad \forall s \geq 0;$

(ii) $G_1(t,s) := \dfrac{\partial G(t,s)}{\partial t} := \begin{cases} -s, & 0 \leq s \leq t, \\ \\ -t, & 0 \leq t \leq s < +\infty; \end{cases}$

(iii) $\displaystyle\lim_{t \to +\infty} \dfrac{G_1(t,s)}{1+t} \in \mathbb{R}, \quad \forall s \geq 0.$

Let $\gamma, \Gamma \in X_{F3}$ such that $\gamma(t) \leq \Gamma(t), \gamma'(t) \leq \Gamma'(t), \forall t \geq 0$ and $\gamma''(+\infty) \leq \Gamma''(+\infty)$. Consider the set

$$E_{F3} = \left\{ (t,x,y,z) \in \mathbb{R}_0^+ \times \mathbb{R}^3 : \begin{array}{c} \gamma(t) \leq x \leq \Gamma(t) \\ \gamma'(t) \leq y \leq \Gamma'(t) \\ \gamma''(+\infty) \leq z(+\infty) \leq \Gamma''(+\infty) \end{array} \right\}.$$

The following Nagumo condition allows some *a priori* bounds on the second derivative of the solution.

**Definition 12.2.3.** A function $f : E_{F3} \to \mathbb{R}$ is said to satisfy a Nagumo-type growth condition in $E_{F3}$ if, for some positive continuous functions $\psi, h$ and some $\nu > 1$, such that

$$\sup \psi(t)(1+t)^\nu < +\infty, \quad \int_0^{+\infty} \frac{s}{h(s)} ds = +\infty, \tag{12.2.4}$$

it verifies

$$|f(t,x,y,z)| \leq \psi(t) h(|z|), \quad \forall (t,x,y,z) \in E_{F3}. \tag{12.2.5}$$

**Lemma 12.2.4.** *Let $f : \mathbb{R}_0^+ \times \mathbb{R}^3 \to \mathbb{R}$ be an $L^1$-Carathéodory function satisfying (12.2.4) and (12.2.5) in $E_{F3}$. Then for every solution $u$ of (12.1.1) satisfying, for $t \geq 0$,*

$$\gamma(t) \leq u(t) \leq \Gamma(t),$$
$$\gamma'(t) \leq u'(t) \leq \Gamma'(t), \tag{12.2.6}$$
$$\gamma''(+\infty) \leq u''(+\infty) \leq \Gamma''(+\infty),$$

*there exists $R > 0$ (not depending on $u$) such that $\|u''\|_2 < R$.*

**Proof.** Let $u$ be a solution of (12.1.1) verifying (12.2.6). Consider $r > 0$ such that

$$r > \max\left\{ |\gamma''(+\infty)|, |\Gamma''(+\infty)| \right\}. \tag{12.2.7}$$

By the previous inequality, $|u''(t)| > r, \forall t \geq 0$ cannot happen because

$$|u''(+\infty)| < r.$$

If $|u''(t)| \leq r, \forall t \geq 0$, taking $R > r$, the proof is complete as

$$\|u''\|_2 = \sup_{t \geq 0} |u''(t)| \leq r < R.$$

In the following, it will be proved that even when there exists $t \geq 0$ such that $|u''(t)| > r$, the norm $\|u''\|_2$ remains bounded.

Suppose there exists $t_0 > 0$ such that $|u''(t_0)| > r$, that is, $u''(t_0) > r$ or $u''(t_0) < -r$.

In the first case, by (12.2.4), one can take $R > r$ such that

$$\int_r^R \frac{s}{h(s)} ds > M \max \left\{ M_1 + \sup_{t \geq 0} \frac{\Gamma'(t)}{1+t} \frac{\nu}{\nu-1}, M_1 - \inf_{t \geq 0} \frac{\gamma'(t)}{1+t} \frac{\nu}{\nu-1} \right\}$$

with $M := \sup_{t \geq 0} \psi(t)(1+t)^\nu$ and $M_1 := \sup_{t \geq 0} \frac{\Gamma'(t)}{(1+t)^\nu} - \inf_{t \geq 0} \frac{\gamma'(t)}{(1+t)^\nu}$.

If condition (12.2.5) holds, then by (12.2.7), there are $t_*, t_+ \geq 0$ such that $t_* < t_+, u''(t_*) = r$ and $u''(t) > r, \forall t \in (t_*, t_+]$. Therefore,

$$\int_{u''(t_*)}^{u''(t_+)} \frac{s}{h(s)} ds = \int_{t_*}^{t_+} \frac{u''(s)}{h(u''(s))} u'''(s) ds \leq \int_{t_*}^{t_+} \psi(s) u''(s) ds$$

$$\leq M \int_{t_*}^{t_+} \frac{u''(s)}{(1+s)^\nu} ds$$

$$= M \int_{t_*}^{t_+} \left[ \left( \frac{u'(s)}{(1+s)^\nu} \right)' + \frac{\nu u'(s)}{(1+s)^{1+\nu}} \right] ds$$

$$\leq M \left( M_1 + \sup_{t \geq 0} \frac{\Gamma'(t)}{1+t} \int_0^{+\infty} \frac{\nu}{(1+s)^\nu} ds \right) < \int_r^R \frac{s}{h(s)} ds.$$

So, $u''(t_+) < R$ and as $t_*$ and $t_+$ are arbitrary in $\mathbb{R}_0^+$, one has that $u''(t) < R, \forall t \geq 0$.

Similarly, the case where there are $t_-, t_* \geq 0$ such that $t_- < t_*$ and $u''(t_*) = -r$, $u''(t) < -r, \forall t \in [t_-, t_*)$ can be proved.

Therefore, $\|u''\|_2 < R, \forall t \geq 0$. $\qquad \square$

The lack of compactness of $X_{F3}$ is overcome by the following lemma which gives a general criterion for relative compactness, suggested in [3] or [51].

**Lemma 12.2.5.** *A set $Z \subset X_{F3}$ is relatively compact if the following conditions hold:*

(i) *all functions from $Z$ are uniformly bounded;*

(ii) *all functions from Z are equicontinuous on any compact interval of* $\mathbb{R}_0^+$;

(iii) *all functions from Z are equiconvergent at infinity, that is, for any given* $\epsilon > 0$, *there exists a* $t_\epsilon > 0$ *such that*

$$\left| \frac{x(t)}{1+t^2} - \lim_{t \to +\infty} \frac{x(t)}{1+t^2} \right| < \epsilon,$$

$$\left| \frac{x'(t)}{1+t} - \lim_{t \to +\infty} \frac{x'(t)}{1+t} \right| < \epsilon,$$

$$\left| x''(t) - \lim_{t \to +\infty} x''(t) \right| < \epsilon \quad \text{for all } t > t_\epsilon, x \in Z.$$

The functions considered as lower and upper solutions for the initial problem are defined as follows with $W^{3,1}\left(\mathbb{R}_0^+\right)$ the usual Sobolev space.

**Definition 12.2.6.** A function $\alpha \in X_{F3} \cap W^{3,1}\left(\mathbb{R}_0^+\right)$ is a lower solution of problem (12.1.1),(12.1.2) if

$$\begin{cases} \alpha'''(t) \geq f(t, \alpha(t), \alpha'(t), \alpha''(t)), & t \geq 0, \\ L_0(\alpha, \alpha(0)) \geq 0, \\ L_1(\alpha, \alpha'(0)) \geq 0, \\ L_2(\alpha, \alpha''(+\infty)) > 0. \end{cases}$$

A function $\beta \in X_{F3} \cap W^{3,1}\left(\mathbb{R}_0^+\right)$ is an upper solution if it satisfies the reverse inequalities.

**Remark 12.2.7.** If $\alpha'(t) \leq \beta'(t)$, a.e. $t \geq 0$ and $\alpha(0) \leq \beta(0)$, by integration on $[0,t]$, one has $\alpha(t) \leq \beta(t), \forall t \geq 0$.

The following lemma, suggested by [141], will ensure the existence and convergence of the derivative of some truncature function to be used forward.

**Lemma 12.2.8** ([141]). *For* $y_1, y_2 \in C^1(\mathbb{R}_0^+)$ *such that* $y_1(t) \leq y_2(t)$, $\forall t \geq 0$, *define*

$$p(t, v) = \begin{cases} y_2(t), & v > y_2(t), \\ v, & y_1(t) \leq v \leq y_2(t), \\ y_1(t), & v < y_1(t). \end{cases}$$

*Then, for each* $v \in C^1\left(\mathbb{R}_0^+\right)$, *the next two properties hold:*

(i) $\frac{d}{dt}p(t, v(t))$ *exists for a.e.* $t \geq 0$;

(ii) *If* $v, v_m \in C^1\left(\mathbb{R}_0^+\right)$ *and* $v_m \to v$ *in* $C^1\left(\mathbb{R}_0^+\right)$, *then*

$$\frac{d}{dt}p(t, v_m(t)) \to \frac{d}{dt}p(t, v(t)) \quad for \ a.e. \ t \geq 0.$$

## 12.3. Existence and localization results

In this section, the existence and localization of at least one solution for the problem (12.1.1),(12.1.2) is proved.

The following assumptions are needed:

($H_1$) There are $\alpha, \beta$ lower and upper solutions of (12.1.1),(12.1.2), respectively, with $\alpha'(t) \leq \beta'(t)$, $t \geq 0$, $\alpha(0) \leq \beta(0)$ and $\alpha''(+\infty) \leq \beta''(+\infty)$;

($H_2$) $f$ satisfies the Nagumo condition on $E_{F3}$ defined with $\gamma = \alpha$ and $\Gamma = \beta$;

$$E_* := \left\{ (t, x, y, z) \in \mathbb{R}_0^+ \times \mathbb{R}^3 : \begin{array}{c} \alpha(t) \leq x \leq \beta(t) \\ \alpha'(t) \leq y \leq \beta'(t) \\ \alpha''(+\infty) \leq z(+\infty) \leq \beta''(+\infty) \end{array} \right\};$$

($H_3$) $f(t, x, y, z)$ verifies the growth condition

$$f(t, \alpha(t), \alpha'(t), \alpha''(t)) \geq f(t, x, \alpha'(t), \alpha''(t)),$$

$$f(t, \beta(t), \beta'(t), \beta''(t)) \leq f(t, x, \beta'(t), \beta''(t))$$

for $t \geq 0$ fixed and $\alpha(t) \leq x \leq \beta(t)$;

($H_4$) The continuous functions $L_i : C(\mathbb{R}_0^+) \times \mathbb{R} \to \mathbb{R}, i = 0, 1, 2$ are such that, for $\alpha \leq v \leq \beta$,

$$\begin{cases} L_i(\alpha, \alpha^{(i)}(0)) \leq L_i(v, \alpha^{(i)}(0)) \text{ and} \\ L_i(\beta, \beta^{(i)}(0)) \geq L_i(v, \beta^{(i)}(0)), \quad \text{for } i = 0, 1; \\ L_2(\alpha, \alpha''(+\infty)) \leq L_2(v, \alpha''(+\infty)) \quad \text{and} \\ L_2(\beta, \beta''(+\infty)) \geq L_2(v, \beta''(+\infty)), \\ \lim_{t \to +\infty} L_2(v, w) \in \mathbb{R}, \quad \text{and } \alpha''(+\infty) \leq w \leq \beta''(+\infty). \end{cases}$$

**Theorem 12.3.1.** *Let* $f : \mathbb{R}_0^+ \times \mathbb{R}^3 \to \mathbb{R}$ *be an* $L^1$-*Carathéodory function. If hypotheses* $(H_1)$–$(H_4)$ *are verified, then problem* (12.1.1),(12.1.2) *has at* *least a solution $u \in X_{F3} \cap W^{3,1}\left(\mathbb{R}_0^+\right)$ and there exists $R > 0$ such that*

$$\alpha(t) \leq u(t) \leq \beta(t), \quad \alpha'(t) \leq u'(t) \leq \beta'(t), \quad -R \leq u''(t) \leq R, \ t \geq 0,$$

*and*

$$\alpha''(+\infty) \leq u''(+\infty) \leq \beta''(+\infty).$$

**Proof.** Let $\alpha, \beta \in X_{F3} \cap W^{3,1}\left(\mathbb{R}_0^+\right)$ verifying $(H_1)$.

Consider the modified and perturbed problem composed by the third-order differential equation

$$u'''(t) = f\left(t, \delta_0(t, u(t)), \delta_1(t, u'(t)), \frac{d}{dt}\left(\delta_1(t, u'(t))\right)\right) \quad (12.3.1)$$

$$+ \frac{1}{1 + t^4} \frac{u'(t) - \delta_1(t, u'(t))}{1 + |u'(t) - \delta_1(t, u'(t))|}, \quad t \geq 0,$$

and the functional boundary equations

$$\begin{cases} u(0) = \delta_0(0, u(0) + L_0\left(\delta_F(u), u(0)\right)), \\ u'(0) = \delta_1(0, u'(0) + L_1\left(\delta_F(u), u'(0)\right)), \\ u''(+\infty) = \delta_\infty(u''(+\infty)) + L_2\left(\delta_F(u), \delta_\infty(u''(+\infty))\right), \end{cases} \quad (12.3.2)$$

where functions $\delta_i : \mathbb{R}_0^+ \times \mathbb{R} \to \mathbb{R}$ are given by

$$\delta_i(t, x) = \begin{cases} \beta^{(i)}(t), & x > \beta^{(i)}(t), \\ x, & \alpha^{(i)}(t) \leq x \leq \beta^{(i)}(t), \quad i = 0, 1, \\ \alpha^{(i)}(t), & x < \alpha^{(i)}(t), \end{cases}$$

$$\delta_\infty(x(+\infty)) = \begin{cases} \beta''(+\infty), & x(+\infty) > \beta''(+\infty), \\ x(+\infty), & \alpha''(+\infty) \leq x(+\infty) \leq \beta''(+\infty), \\ \alpha''(+\infty), & x(+\infty) < \alpha''(+\infty), \end{cases}$$

$$\delta_F(v) = \begin{cases} \beta, & v > \beta, \\ v, & \alpha \leq v \leq \beta, \\ \alpha, & v < \alpha. \end{cases}$$

For clarity, the proof follows several steps:

**Step 1.** *If $u$ is a solution of (12.3.1),(12.3.2), then*

$$\alpha'(t) \leq u'(t) \leq \beta'(t), \ \alpha(t) \leq u(t) \leq \beta(t), \ -R \leq u''(t) \leq R, \ \forall t \geq 0,$$

and

$$\alpha''(+\infty) \le u''(+\infty) \le \beta''(+\infty).$$

Let $u$ be a solution of the modified problem (12.3.1),(12.3.2) and suppose, by contradiction, that there exists $t \ge 0$ such that $\alpha'(t) > u'(t)$. Therefore,

$$\inf_{t \ge 0}(u'(t) - \alpha'(t)) < 0.$$

- If the infimum is attained at $t = 0$, then

$$\min_{t \ge 0}(u'(t) - \alpha'(t)) = u'(0) - \alpha'(0) < 0,$$

therefore, the contradiction holds

$$0 > u'(0) - \alpha'(0) = \delta_1(0, u'(0) + L_1\left(\delta_F(u), u'(0)\right)) - \alpha'(0)$$

$$\ge \alpha'(0) - \alpha'(0) = 0.$$

- If the infimum occurs at $t = +\infty$, then

$$\inf_{t \ge 0}(u'(t) - \alpha'(t)) = u'(+\infty) - \alpha'(+\infty) < 0.$$

Therefore, $u''(+\infty) - \alpha''(+\infty) \le 0$ and by $(H_4)$ and Definition 12.2.6, the contradiction holds:

$$0 \ge u''(+\infty) - \alpha''(+\infty) = \delta_\infty(u''(+\infty)) + L_2\left(\delta_F(u), \delta_\infty(u''(+\infty))\right)$$

$$\ge L_2(\delta_F(u), \alpha''(+\infty)) \ge L_2(\alpha, \alpha''(+\infty)) > 0. \tag{12.3.3}$$

- If there is an interior point $t_* \in \mathbb{R}^+$ such that

$$\min_{t \ge 0}(u'(t) - \alpha'(t)) := u'(t_*) - \alpha'(t_*) < 0,$$

then there exists $0 \le t_1 < t_*$ where

$$u'(t) - \alpha'(t) < 0, \quad u''(t) - \alpha''(t) \le 0, \quad \forall t \in [t_1, t_*],$$

$$u'''(t) - \alpha'''(t) \ge 0, \quad \text{a.e. } t \in [t_1, t_*].$$

Therefore, for $t \in [t_1, t_*]$ by $(H_3)$ and Definition 12.2.6, the contradiction holds:

$$0 \le \int_{t_1}^{t} [u'''(s) - \alpha'''(s)] \, ds$$

$$= \int_{t_1}^{t} \left[ f\left( (s, \delta_0(s, u(s)), \delta_1(s, u'(s)), \frac{d}{ds}(\delta_1(s, u'(s)))\right) \right.$$

$$\left. + \frac{1}{1 + s^4} \frac{u'(s) - \delta_1(s, u'(s))}{1 + |u'(s) - \delta_1(s, u'(s))|} - \alpha'''(s) \right] ds$$

$$\leq \int_{t_1}^{t} \left[ f(s, \alpha(s), \alpha'(s), \alpha''(s)) + \frac{u'(s) - \alpha'(s)}{1 + |u'(s) - \alpha'(s)|} - \alpha'''(s) \right] ds$$

$$\leq \int_{t_1}^{t} \left[ \frac{u'(s) - \alpha'(s)}{1 + |u'(s) - \alpha'(s)|} \right] ds < 0.$$

So, $u'(t) \geq \alpha'(t)$ for $t > 0$.

In a similar way, it can be proved that $u'(t) \leq \beta'(t)$, and, therefore,

$$\alpha'(t) \leq u'(t) \leq \beta'(t), \quad \forall t \geq 0. \tag{12.3.4}$$

Remark that $\alpha(0) \leq u(0)$, otherwise, by $(H_4)$ and Definition 12.2.6, the contradiction will happen:

$$0 > u(0) - \alpha(0) = \delta_0(0, u(0) + L_0(\delta_F(u), u(0))) - \alpha(0)$$

$$\geq L_0(\delta_F(u), u(0))) \geq L_0(\alpha, \alpha(0))) \geq 0.$$

Analogously, it can be proved that $u(0) \leq \beta(0)$. So, integrating (12.3.4) in $[0, t]$, it is easily obtained that $\alpha(t) \leq u(t) \leq \beta(t), \forall t \geq 0$.

Arguing like in (12.3.3), one can prove that $u''(+\infty) \geq \alpha''(+\infty)$ and, similarly, that $u''(+\infty) \leq \beta''(+\infty)$.

Therefore, $(t, u(t), u'(t), u''(t)) \in E_*$ and the inequality $-R \leq u''(t) \leq R$ is a direct consequence of Lemma 12.2.4.

**Step 2.** *The problem* (12.3.1),(12.3.2) *has at least one solution.*

Define the operator $T : X_{F3} \to X_{F3}$

$$Tu(t) = \Delta + \Gamma t + \frac{\Psi t^2}{2} + \int_{0}^{+\infty} G(t, s) F_u(s) ds,$$

where

$$\Delta := \delta_0\left(0, u(0) + L_0(\delta_F(u), u(0))\right),$$

$$\Gamma := \delta_1(0, u'(0) + L_0(\delta_F(u), u'(0))),$$

$$\Psi := \delta_\infty(u''(+\infty)) + L_2\left(\delta_F(u), \delta_\infty(u''(+\infty))\right),$$

$G(t, s)$ is the Green function given by (12.2.3) associated with the problem

$$\begin{cases} u'''(t) = F_u(t), \quad t \geq 0, \\ u(0) = \Delta, \\ u'(0) = \Gamma, \\ u''(+\infty) = \Psi, \end{cases} \tag{12.3.5}$$

and

$$F_u(t) := f\left(t, \delta_0(t, u(t)), \delta_1(t, u'(t)), \frac{d}{dt}(\delta_1(t, u'(t)))\right)$$

$$+ \frac{1}{1 + t^4} \frac{u'(t) - \delta_1(t, u'(t))}{1 + |u'(t) - \delta_1(t, u'(t))|}.$$

By Lemma 12.2.1, the fixed points of $T$ are solutions of (12.3.5) and, therefore, of problem (12.3.1),(12.3.2).

So, it is enough to prove that $T$ has a fixed point.

**Step 2.1.** $T$ *is well defined and, for a compact* $D \subset X_{F3}$, $TD \subset D$.

As $f$ is an $L^1$-Carathéodory function, $Tu \in C^2\left(\mathbb{R}_0^+\right)$ and for any $u \in X_{F3}$ with

$$\rho > \max\{\|u\|_{X_{F3}}, \|\alpha\|_{X_{F3}}, \|\beta\|_{X_{F3}}, R\},$$

there exists a positive function $\phi_\rho(t)$ such that $t^2\phi_\rho(t), t\phi_\rho(t), \phi_\rho(t) \in L^1\left(\mathbb{R}_0^+\right)$ and

$$\int_0^{+\infty} |F_u(s)|\, ds \le \int_0^{+\infty} \left(\phi_\rho(s) + \frac{1}{1 + s^4}\right) ds < +\infty,$$

$$\int_0^{+\infty} |sF_u(s)|\, ds \le \int_0^{+\infty} \left(s\phi_\rho(s) + \frac{s}{1 + s^4}\right) ds < +\infty,$$

$$\int_0^{+\infty} |s^2 F_u(s)|\, ds \le \int_0^{+\infty} \left(s^2\phi_\rho(s) + \frac{s^2}{1 + s^4}\right) ds < +\infty,$$

that is, $F_u, tF_u, t^2F_u \in L^1\left(\mathbb{R}_0^+\right)$.

By Lebesgue Dominated Convergence Theorem, Lemma 12.2.3 and $(H_4)$, setting

$$\lim_{t \to +\infty} L_2\left(\delta_F(u), \delta_\infty(u''(+\infty))\right) := L,$$

and

$$M_\infty := \max\left\{|\alpha''(+\infty)| + |L|, |\beta''(+\infty)| + |L|\right\},$$

one has

$$\lim_{t \to +\infty} \frac{(Tu)(t)}{1 + t^2} = \lim_{t \to +\infty} \frac{\Delta + \Gamma t + \frac{\Psi t^2}{2}}{1 + t^2} + \int_0^{+\infty} \lim_{t \to +\infty} \frac{G(t, s)}{1 + t^2} F_u(s)\, ds$$

$$\le \frac{M_\infty}{2} + \frac{1}{2}\int_0^{+\infty} \left(\phi_\rho(s) + \frac{1}{1 + s^4}\right) ds < +\infty,$$

$$\lim_{t \to +\infty} \frac{(Tu)'(t)}{1+t} = \lim_{t \to +\infty} \frac{\Gamma + \Psi t}{1+t} + \int_0^{+\infty} \lim_{t \to +\infty} \frac{G_1(t,s)}{1+t} F_u(s) ds$$

$$\leq M_\infty + \int_0^{+\infty} \left( \phi_\rho(s) + \frac{1}{1+s^4} \right) ds < +\infty,$$

$$\lim_{t \to +\infty} (Tu)''(t) = M_\infty + \lim_{t \to +\infty} \int_t^{+\infty} F_u(s) ds < +\infty.$$

Therefore, $Tu \in X_{F3}$.

Consider now the subset $D \subset X_{F3}$ given by

$$D := \{ x \in X_{F3} : \|u\|_{X_{F3}} < \rho_0 \},$$

with $\rho_0 > 0$, such that

$$\rho_0 > \max \{ |\alpha(0)|, |\beta(0)| \} + \max \{ |\alpha'(0)|, |\beta'(0)| \} + |k_0|$$

$$+ \int_0^{+\infty} M(s) \left( \phi_\rho(s) + \frac{1}{1+s^4} \right) ds,$$

where

$$k_0 := \max \{ |\alpha''(+\infty)|, |\beta''(+\infty)| \} + \sup_{t \geq 0} L_2(v, w),$$

for $\alpha \leq v \leq \beta$ and $\alpha''(+\infty) \leq w \leq \beta''(+\infty)$, and

$$M(s) := \max \left\{ \sup_{t \geq 0} \frac{|G(t,s)|}{1+t^2}, \sup_{t \geq 0} \frac{|G_1(t,s)|}{1+t}, 1 \right\}.$$

So, for $t \geq 0$,

$$\|Tu\|_0 = \sup_{t \geq 0} \frac{|Tu(t)|}{1+t^2}$$

$$\leq \sup_{t \geq 0} \left( \frac{\left| \Delta + \Gamma t + \frac{\Psi t^2}{2} \right|}{1+t^2} \right) + \sup_{t \geq 0} \left( \int_0^{+\infty} \frac{|G(t,s)|}{1+t^2} |F_u(s)| ds \right)$$

$$\leq |\Delta| + |\Gamma| + \frac{|\Psi|}{2} + \int_0^{+\infty} M(s) \left( \phi_{\rho_0}(s) + \frac{1}{1+s^4} \right) ds < \rho_0,$$

$$\|(Tu)'\|_1 = \sup_{t \geq 0} \frac{|(Tu)'|}{1+t} \leq \sup_{t \geq 0} \left( \frac{|\Gamma + \Psi t|}{1+t} + \int_0^{+\infty} \frac{|G_1(t,s)|}{1+t} |F_u(s)| ds \right)$$

$$\leq |\Gamma| + |\Psi| + \int_0^{+\infty} M(s) \left( \phi_{r_1}(s) + \frac{1}{1+s^4} \right) ds < \rho_0,$$

and

$$\|(Tu)''\|_2 = \sup_{t \geq 0} |(Tu)''| \leq \sup_{t \geq 0} \left( |\Psi| + \int_t^{+\infty} |F_u(s)| \, ds \right)$$

$$\leq \sup_{t \geq 0} \left( |\Psi| + \int_t^{+\infty} \phi_{r_1}(s) + \frac{1}{1+s^4} ds \right) < \rho_0.$$

So, $TD \subset D$.

**Step 2.2.** *T is continuous.*

Consider a convergent sequence $u_n \to u$ in $X_{F3}$, there exists $\rho_1 > 0$ such that $\max\{\sup_n \|u_n\|_{X_{F3}}, \|\alpha\|_{X_{F3}}, \|\beta\|_{X_{F3}}, R\} < \rho_1$. By Lemma 12.2.8, one has

$$\|Tu_n - Tu\|_X = \max \{ \|Tu_n - Tu\|_0, \|(Tu_n)' - (Tu)'\|_1,$$

$$\|(Tu_n)'' - (Tu)''\|_2 \}$$

$$\leq \int_0^{+\infty} M(s) |F_{u_n}(s) - F_u(s)| \, ds \longrightarrow 0, \text{ as } n \to +\infty.$$

**Step 2.3.** *T is compact.*

Let $B \subset X_{F3}$ be any bounded subset. Therefore, there is $r > 0$ such that $\|u\|_{X_{F3}} < r, \forall u \in B$.

For each $u \in B$, and for $\max\{r, R, \|\alpha\|_{X_{F3}}, \|\beta\|_{X_{F3}}\} < r_1$, similar arguments to Step 2.1 can be applied to prove that $\|Tu\|_0$, $\|(Tu)'\|_1$ and $Vert(Tu)''\|_2$ are finite.

So, $\|Tu\|_{X_{F3}} = \max\{\|Tu\|_0, \|(Tu)'\|_1, \|(Tu)''\|_2\} < +\infty$, that is, $TB$ is uniformly bounded in $X_{F3}$.

$TB$ is equicontinuous because, for $L > 0$ and $t_1, t_2 \in [0, L]$, one has, as $t_1 \to t_2$,

$$\left| \frac{Tu(t_1)}{1+t_1^2} - \frac{Tu(t_2)}{1+t_2^2} \right|$$

$$\leq \left| \frac{\Delta + \Gamma t_1 + \frac{\Psi t_1}{2}}{1+t_1^2} - \frac{\Delta + \Gamma t_2 + \frac{\Psi t_2}{2}}{1+t_2^2} \right|$$

$$+ \int_0^{+\infty} \left| \frac{G(t_1, s)}{1+t_1^2} - \frac{G(t_2, s)}{1+t_2^2} \right| |F(u(s))| \, ds$$

$$\leq \left| \frac{\Delta + \Gamma t_1 + \frac{\Psi t_1}{2}}{1 + t_1^2} - \frac{\Delta + \Gamma t_2 + \frac{\Psi t_2}{2}}{1 + t_2^2} \right|$$

$$+ \int_0^{+\infty} \left| \frac{G(t_1, s)}{1 + t_1^2} - \frac{G(t_2, s)}{1 + t_2^2} \right| \left( \phi_{r_1}(s) + \frac{1}{1 + s^4} \right) ds \longrightarrow 0,$$

$$\left| \frac{(Tu)'(t_1)}{1 + t_1} - \frac{(Tu)'(t_2)}{1 + t_2} \right|$$

$$\leq \left| \frac{\Gamma + \Psi t_1}{1 + t_1} - \frac{\Gamma + \Psi t_2}{1 + t_2} \right|$$

$$+ \int_0^{+\infty} \left| \frac{G_1(t_1, s)}{1 + t_1} - \frac{G_1(t_2, s)}{1 + t_2} \right| |F(u(s))| \, ds$$

$$\leq \left| \frac{\Gamma + \Psi t_1}{1 + t_1} - \frac{\Gamma + \Psi t_2}{1 + t_2} \right|$$

$$+ \int_0^{+\infty} \left| \frac{G_1(t_1, s)}{1 + t_1} - \frac{G_1(t_2, s)}{1 + t_2} \right| \left( \phi_{r_1}(s) + \frac{1}{1 + s^4} \right) ds \longrightarrow 0,$$

$$|(Tu)''(t_1) - (Tu)''(t_2)| = \left| \int_{t_1}^{+\infty} F_u(s) ds - \int_{t_2}^{+\infty} F_u(s) ds \right|$$

$$\leq \int_{t_1}^{t_2} |F_u(s)| \, ds$$

$$\leq \int_{t_1}^{t_2} \left( \phi_{r_1}(s) + \frac{1}{1 + s^4} \right) ds \longrightarrow 0.$$

Moreover, $TB$ is equiconvergent at infinity because, as $t \to +\infty$,

$$\left| \frac{Tu(t)}{1 + t^2} - \lim_{t \to +\infty} \frac{Tu(t)}{1 + t^2} \right|$$

$$\leq \left| \frac{\Delta + \Gamma t + \frac{\Psi t^2}{2}}{1 + t^2} - \frac{\Psi}{2} \right|$$

$$+ \int_0^{+\infty} \left| \frac{G(t, s)}{1 + t^2} + \frac{1}{2} \right| |F_u(s)| \, ds$$

$$\leq \left| \frac{\Delta + \Gamma t + \frac{\Psi t^2}{2}}{1 + t^2} - \frac{\Psi}{2} \right|$$

$$+ \int_0^{+\infty} \left| \frac{G(t,s)}{1+t^2} + \frac{1}{2} \right| \left( \phi_{\rho_1} + \frac{1}{1+s^4} \right) ds \to 0,$$

$$\left| \frac{(Tu)'(t)}{1+t} - \lim_{t \to +\infty} \frac{Tu(t)}{1+t} \right|$$

$$\leq \left| \frac{\Gamma + \Psi t}{1+t} - \Psi \right|$$

$$+ \int_0^{+\infty} \left| \frac{G_1(t,s)}{1+t} + 1 \right| |F_u(s)| \, ds$$

$$\leq \left| \frac{\Gamma + \Psi t}{1+t} - \Psi \right|$$

$$+ \int_0^{+\infty} \left| \frac{G_1(t,s)}{1+t} + 1 \right| \left( \phi_{\rho_1} + \frac{1}{1+s^4} \right) ds \to 0,$$

and

$$\left| (Tu)''(t) - \lim_{t \to +\infty} (Tu)''(t) \right| = \int_t^{+\infty} |F_u(s)| \, ds$$

$$\leq \int_t^{+\infty} \left( \phi_{\rho_1} + \frac{1}{1+s^4} \right) ds \longrightarrow 0.$$

So, by Lemma 12.2.5, $TB$ is relatively compact.

Then by Schauder's fixed-point theorem (Theorem 1.2.6), $T$ has at least one fixed point $u_1 \in X_{F3}$.

**Step 3.** $u_1$ *is a solution of* (12.1.1), (12.1.2).

Suppose, by contradiction, that

$$\alpha(0) > u_1(0) + L_0(\delta_F, u_1(0)).$$

Then, by (12.3.2), $u_1(0) = \alpha(0)$ and, by $(H_4)$ and Definition 12.2.6, the following contradiction holds:

$$u_1(0) + L_0(\delta_F(u_1), u_1(0)) = \alpha(0) + L_0(\delta_F(u_1), \alpha(0))$$

$$\geq \alpha(0) + L_0(\alpha, \alpha(0)) \geq \alpha(0).$$

So, $\alpha(0) \leq u_1(0) + L_0(\delta_F, u_1(0))$. In a similar way, it can be proved that $u_1(0) + L_0(\delta_F(u_1), u_1(0)) \leq \beta(0)$.

Assuming, by contradiction, that $\alpha'(0) > u_1'(0) + L_1(\delta_F(u_1), u_1'(0))$, $u_1'(0) = \alpha'(0)$ and, by $(H_4)$ and Definition 12.2.6, the following contradiction is achieved:

$$u_1'(0) + L_1(\delta_F(u_1), u_1'(0)) = \alpha'(0) + L_1(\delta_F(u_1), \alpha'(0))$$
$$\geq \alpha'(0) + L_1(\alpha, \alpha'(0)) \geq \alpha'(0).$$

So, $\alpha'(0) \leq u_1'(0) + L_1(\delta_F(u_1), u_1'(0))$. By similar arguments, it can be proved that $u_1'(0) + L_1(\delta_F(u_1), u_1'(0)) \leq \beta'(0)$.

By Step 1, $\alpha(0) \leq u_1(0) \leq \beta(0), \alpha'(0) \leq u_1'(0) \leq \beta'(0)$ and $-R \leq u_1''(+\infty) \leq R$, and therefore, $u_1(t)$ verifies the differential equation (12.1.1) and boundary conditions (12.1.2), that is, $u_1$ is a solution of (12.1.1), (12.1.2). $\qquad\square$

## 12.4. Falkner–Skan equation

A classical third-order differential equation, known as the Falkner–Skan equation, is of the form

$$u'''(t) + au(t)u''(t) + b(1 - (u'(t))^2) = 0, \quad t \geq 0. \tag{12.4.1}$$

This general equation is obtained from partial differential equations by using some transformation technique (see [155]).

When $b = 0$, equation (12.4.1) is known as the Blasius equation, and it models the behavior of a viscous flow over a flat plate. A boundary layer is created by a two-dimensional flow over a fixed impenetrable surface, and particles move more slowly near the surface than near the free stream. In this way, equation (12.4.1) can be subject to the following boundary conditions on the half line:

$$u(0) = 0, \quad u'(0) = 0, \quad u'(+\infty) = 1. \tag{12.4.2}$$

In the literature, only numerical techniques are applied to deal with these types of problems, (12.4.1),(12.4.2), with general $a, b$ (see, for instance, [157]).

To illustrate the main result, let us consider a boundary value problem of this family, composed by the third-order fully differential equation

$$u'''(t) = \frac{(u'(t))^2 - 1}{1 + t^6} - \frac{u(t)|u''(t)|}{e^{3t}} + \frac{u''(t)}{1 + t^4}, \quad t \geq 0, \tag{12.4.3}$$

and the functional boundary conditions on the half-line:

$$\int_0^{+\infty} \frac{|u(t)|}{(t^2 + t + 1)(t^2 + 1)} dt - 2u(0) = 0,$$

$$u'(0) = 1, \qquad (12.4.4)$$

$$\inf_{t \geq 0} \frac{u(t)}{1 + t^2} - u''(+\infty) = -0.5.$$

Remark that the above problem is a particular case of (12.1.1),(12.1.2) with

$$f(t, x, y, z) = \frac{y^2 - 1}{1 + t^6} - \frac{x|z|}{e^{3t}} + \frac{z}{1 + t^4},$$

$$L_0(a, b) = \int_0^{+\infty} \frac{|a(t)|}{(t^2 + t + 1)(t^2 + 1)} dt - 2b,$$

$$L_1(a, c) = c - 1, \qquad (12.4.5)$$

$$L_2(a, d) = \inf_{t \geq 0} \frac{a(t)}{1 + t^2} - d + 0.5.$$

Functions $\beta(t) = t^2 + t + 1$ and $\alpha(t) = t$ are, respectively, upper and lower solutions of the problem (12.4.3),(12.4.4), verifying $(H_1)$.

The nonlinear function $f : \mathbb{R}_0^+ \times \mathbb{R}^3 \to \mathbb{R}$ verifies the assumptions of Theorem 12.3.1. In fact,

- $f$ is an $L^1$-Carathéodory function as for $|x| < \rho(1 + t^2)$, $|y| < \rho(1 + t)$ and $|z| < \rho$, one has

$$|f(t, x, y, z)| \leq \frac{\rho^2(1 + t)^2 + 1}{1 + t^6} + \frac{\rho^2(1 + t^2)}{e^{3t}} + \frac{\rho}{1 + t^4} := \phi_\rho(t)$$

with $\phi_\rho, t\phi_\rho, t^2\phi_\rho \in L^1(\mathbb{R}_0^+)$;
- $f$ verifies the Nagumo condition on the set

$$E_* = \left\{ (t, x, y, z) \in \mathbb{R}_0^+ \times \mathbb{R}^3 : \begin{array}{c} t \leq x \leq t^2 + t + 1 \\ 1 \leq y \leq 2t + 1 \\ 0 \leq z(+\infty) \leq 2 \end{array} \right\}$$

with $\psi(t) = \frac{k}{1 + t^4}$ and $h = 1$, where $k > 0$ is a real constant;
- $f(t, x, y, z)$ is nonincreasing in $x$, therefore, it satisfies $(H_3)$;

As the functions $L_i$, $i = 0, 1, 2$, given by (12.4.5) verify $(H_4)$, then, by Theorem 12.3.1, there is at least a solution $u$ of (12.4.3),(12.4.4) such that

$$t \leq u(t) \leq t^2 + t + 1, \ 1 \leq u'(t) \leq 2t + 1, \ 0 \leq u''(t) \leq 2, \ \text{for } t \geq 0.$$

This localization part shows that this solution is unbounded, nonnegative, increasing and convex.

Chapter 13

# φ-Laplacian Equations with Functional Boundary Conditions

## 13.1. Introduction

This chapter is concerned with the study of $\phi$-Laplacian equations, sometimes called in the literature as half-linear equations. More precisely, we consider a fully nonlinear equation on the half-line

$$(\phi(u'(t)))' + q(t)f(t, u(t), u'(t)) = 0, \quad t \geq 0, \tag{13.1.1}$$

where $\phi : \mathbb{R} \to \mathbb{R}$ is an increasing homeomorphism with $\phi(0) = 0$, $f : \mathbb{R}_0^+ \times \mathbb{R}^2 \to \mathbb{R}$ and $q : \mathbb{R}^+ \to \mathbb{R}_0^+$ are both continuous functions, verifying adequate assumptions, but $q$ is allowed to have a singularity when $t = 0$, coupled with the functional boundary conditions

$$L(u, u(0), u'(0)) = 0, \quad u'(+\infty) := \lim_{t \to +\infty} u'(t) = B, \tag{13.1.2}$$

where $L : C\left(\mathbb{R}_0^+\right) \times \mathbb{R}^2 \to \mathbb{R}$ is a continuous function with properties to be expressed later and $B \in \mathbb{R}$.

Boundary value problems, usually, are considered on compact domains. However, problems on the half-line are becoming increasingly more popular in the literature due to their applications in fields like engineering, chemistry and biology (see, for instance, [117, 147, 151]). Moreover, if equation (13.1.1) is considered on the whole real line, some techniques to guarantee the existence of homoclinic and heteroclinic solutions have been developed in the recent years, as it can be seen in [110–112, 115].

Problems defined on unbounded domains require more delicate procedures to deal with the lack of compactness. In this chapter, this is overcome by applying the so-called Bielecki norm and the equiconvergence at $\infty$, as in [51].

It is important to note that, in this chapter, two types of new features are introduced:

- The homeomorphism $\phi$ does not need to be surjective, that is $\phi(\mathbb{R})$ can be different from $\mathbb{R}$. This is overcome by an auxiliary surjective homeomorphism that extends, eventually, $\phi$.
- A new and more general type of boundary conditions, given by a functional that can depend globally on the unknown function.

Moreover, this method can be applied to classical or singular $\phi$-Laplacian, that is, even for homeomorphism $\phi : (-a, a) \to \mathbb{R}$, with $0 < a < +\infty$ (for more details, see [24, 37]).

In general, the lower and upper solutions method is a very adequate and useful technique to deal with functional boundary value problems as it provides not only the existence of bounded or unbounded solutions, but also their localization and, from that, some qualitative data about solutions, their variation and behavior (see [35, 71, 72, 99, 100, 113]).

The technique used in this chapter follows the work [68], and applies some arguments suggested in [57], combined with the upper and lower solutions and a Nagumo condition to control the first derivative. The usage of such tool allows to improve the existent solutions, namely the introduction of functional boundary conditions in the problem. These boundary conditions are very general in nature. Not only they generalize most of the classical boundary conditions, but they also cover the separated and multipoint cases, nonlocal or integral conditions or other boundary conditions with maximum/minimum arguments, that is, for example, of the type

$$u(0) = \max_{t \geq 0} u(t) \text{ or } u'(\tau) = \min_{t \geq 0} u'(t), \text{ with } \tau \geq 0,$$

provided that the assumptions on $L$ are satisfied.

The chapter is organized as follows. In the first section, some auxiliary result are defined such as the space, the weighted norms, lower and upper solutions to be used and the necessary lemmas to proceed. The second section contains new results of existence and localization of solutions. Finally, two examples, which are not covered by the existent literature, show the applicability of the main theorems. In the first one, the Nagumo conditions are verified. On the other hand, in the second one, these assumptions are replaced by a stronger condition on lower and upper solutions together with a local monotone growth on $f$.

## 13.2. Preliminary results

In this section, some definitions and auxiliary results needed for the proof of the main result are presented. Consider the following space

$$X_\phi = \left\{ x \in C^1(\mathbb{R}_0^+) : \lim_{t \to +\infty} \frac{x(t)}{e^{\theta t}} \in \mathbb{R} \right\}$$

equipped with a Bielecki norm type in $C^1(\mathbb{R}_0^+)$,

$$\|x\|_{X_\phi} := \max\left\{ \|x\|_0, \|x'\|_1 \right\},$$

where

$$\|w\|_0 = \sup_{t \geq 0} \frac{|w(t)|}{e^{\theta t}} \quad \text{and} \quad \|w\|_1 = \sup_{t \geq 0} |w(t)|.$$

In this way, it is clear that $(X_\phi, \|\cdot\|_{X_\phi})$ is a Banach space.

In addition, the following conditions must hold:

(H1) $\phi : \mathbb{R} \to \mathbb{R}$ is an increasing homeomorphism with $\phi(0) = 0$.

(H2) The function $f : \mathbb{R}^+ \times \mathbb{R}^2 \to \mathbb{R}$ is continuous and $f(t, x, y)$ is uniformly bounded for $t > 0$ when $x$ and $y$ are bounded.

(H3) The function $q : \mathbb{R}^+ \to \mathbb{R}_0^+$ is integrable, not identically to 0 in a subinterval of $\mathbb{R}^+$.

(H4) $L : C(\mathbb{R}^+) \times \mathbb{R}^2 \to \mathbb{R}$ is a continuous function, nondecreasing in the first and third variables.

The approach to problem (13.1.1),(13.1.2) will be from the perspective of a fixed point problem. In this order, the next lemmas will establish the link between problem (13.1.1),(13.1.2) and its integral formulation.

Let $\gamma, \Gamma \in X_\phi$ be such that $\gamma(t) \leq \Gamma(t), \forall t \geq 0$. Consider the set, for $\theta > 0$,

$$E_\theta = \left\{ (t, x, y) \in \mathbb{R}_0^+ \times \mathbb{R}^2 : \frac{\gamma(t)}{e^{\theta t}} \leq x \leq \frac{\Gamma(t)}{e^{\theta t}} \right\}.$$

The following Nagumo condition allows some *a priori* bounds on the first derivative of the solution.

**Definition 13.2.1.** A function $f : E_\theta \to \mathbb{R}$ is said to satisfy a Nagumo-type growth condition in $E_\theta$ if, for some positive and continuous functions $\psi, h$, such that

$$\sup_{t \geq 0} \psi(t) < +\infty, \quad \int_0^{+\infty} \frac{|\phi^{-1}(s)|}{h(|\phi^{-1}(s)|)} ds = +\infty, \tag{13.2.1}$$

it verifies

$$|q(t)f(t,x,y)| \leq \psi(t)h(|y|), \quad \forall(t,x,y) \in E_\theta. \tag{13.2.2}$$

**Lemma 13.2.2.** *Let* $f : \mathbb{R}_0^+ \times \mathbb{R}^2 \to \mathbb{R}$ *be a continuous function satisfying a Nagumo-type growth condition in* $E_\theta$. *Then there exists* $N > 0$ (*not depending on* $u$) *such that every solution* $u$ *of* (13.1.1),(13.1.2) *with*

$$\frac{\gamma(t)}{e^{\theta t}} \leq u(t) \leq \frac{\Gamma(t)}{e^{\theta t}}, \quad for \ t \geq 0, \ \theta > 0,$$

*has*

$$\|u'\|_1 < N. \tag{13.2.3}$$

**Proof.** Let $u$ be a solution of (13.1.1),(13.1.2) with $(t, u(t), u'(t)) \in E_\theta$. Consider $r > 0$, such that

$$r > |B|. \tag{13.2.4}$$

If $|u'(t)| \leq r, \forall t \geq 0$, taking $N > r$ the proof is complete as

$$\|u'\|_1 = \sup_{t \geq 0} |u'(t)| \leq r < N.$$

Suppose there exists $t_0 \geq 0$ such that $|u'(t_0)| > N$, that is, $u'(t_0) > N$ or $u'(t_0) < -N$.

In the first case, by (13.2.1), one can take $N > r$ such that

$$\int_{\phi(r)}^{\phi(N)} \frac{|\phi^{-1}(s)|}{h(|\phi^{-1}(s)|)} ds > M \left( \sup_{t \geq 0} \frac{\Gamma(t)}{e^{\theta t}} - \inf_{t \geq 0} \frac{\gamma(t)}{e^{\theta t}} \right) \tag{13.2.5}$$

with $M := \sup_{t \geq 0} \psi(t)$.

Consider $t_1, t_2 \in [t_0, +\infty)$ such that $t_1 < t_2$, $u'(t_1) = N$, $u'(t_2) = r$ and $r \leq u'(t) \leq N, \forall t \in [t_1, t_2]$. Therefore, the following contradiction with (13.2.5) is achieved:

$$\int_{\phi(r)}^{\phi(N)} \frac{|\phi^{-1}(s)|}{h(|\phi^{-1}(s)|)} ds = \int_{\phi(u'(t_2))}^{\phi(u'(t_1))} \frac{\phi^{-1}(s)}{h(\phi^{-1}(s))} ds$$

$$= \int_{t_2}^{t_1} \frac{u'(s)}{h(u'(s))} (\phi(u'(s)))' ds$$

$$= -\int_{t_1}^{t_2} \frac{q(s)f(s, u(s), u'(s))}{h(u'(s))} u'(s) ds$$

$$\leq \int_{t_1}^{t_2} \frac{|q(s)f(s,u(s),u'(s))|}{h(u'(s))} \, u'(s) \, ds$$

$$\leq \int_{t_1}^{t_2} \psi(s)u'(s) \, ds \leq M \int_{t_1}^{t_2} u'(s) \, ds$$

$$\leq M(u(t_2) - u(t_1)) \leq M \left( \sup_{t \geq 0} \frac{\Gamma(t)}{e^{\theta t}} - \inf_{t \geq 0} \frac{\gamma(t)}{e^{\theta t}} \right).$$

So $u'(t) < N, \forall t \geq 0$.

Similarly, it can be proved that $u'(t) > -N, \forall t \geq 0$, and, therefore, $\|u'\|_1 < N, \forall t \geq 0$. $\qquad \square$

Define a surjective homeomorphism $\varphi : \mathbb{R} \to \mathbb{R}$ as

$$\varphi(y) = \begin{cases} \phi(y) & \text{if } |y| \leq R \\ \dfrac{\phi(R) - \phi(-R)}{2R} y + \dfrac{\phi(R) + \phi(-R)}{2} & \text{if } |y| > R \end{cases} \qquad (13.2.6)$$

with $R > 0$ is to be defined later.

**Lemma 13.2.3.** *Let* $v \in L^1(\mathbb{R}_0^+)$. *Then* $u \in X_\phi$ *such that* $(\varphi(u'(t))) \in AC(\mathbb{R}_0^+)$ *is the unique solution of*

$$(\varphi(u'(t)))' + v(t) = 0, \quad t \geq 0 \qquad (13.2.7)$$

$$u(0) = A$$

$$u'(+\infty) = B,$$

*with* $A, B \in \mathbb{R}$, *if and only if*

$$u(t) = A + \int_0^t \varphi^{-1} \left( \varphi(B) + \int_s^{+\infty} v(\tau) \, d\tau \right) ds \qquad (13.2.8)$$

**Proof.** Let $u \in X_\phi$ be a solution of (13.2.7). Then

$$(\varphi(u'(t)))' = -v(t),$$

and by integration, one has

$$\varphi(u'(t)) = \varphi(B) + \int_t^{+\infty} v(s) ds.$$

As $\varphi$ is continuous and $\varphi(\mathbb{R}) = \mathbb{R}$, then

$$u'(t) = \varphi^{-1} \left( \varphi(B) + \int_t^{+\infty} v(s) ds \right)$$

and by integration again,

$$u(t) = A + \int_0^t \varphi^{-1}\left(\varphi(B) + \int_s^{+\infty} v(\tau)\,d\tau\right)ds.$$

$\square$

The lack of compactness is overcome by the following lemma, which will provide a general criteria for relative compactness.

**Lemma 13.2.4.** ([51]). *Let $M \subset X_\phi$. The set $M$ is said to be relatively compact if the following conditions hold:*

(a) *$M$ is uniformly bounded in $X_\phi$;*
(b) *the functions belonging to $M$ are equicontinuous on any compact interval of $\mathbb{R}_0^+$;*
(c) *the functions $f$ from $M$ are equiconvergent at $+\infty$, i.e., given $\varepsilon > 0$, there exists $T(\varepsilon) > 0$ such that $\|f(t) - f(+\infty)\|_{X_\phi} < \varepsilon$ for any $t > T(\varepsilon)$ and $f \in M$.*

The adaptation of the Euclidean norm of $\mathbb{R}^n$ to the weighted norms of $X_\phi$ is a scholar exercise and, by this reason, was omitted.

To prove the main result, it is important to rely on the upper and lower solutions method. The functions to be considered as upper and lower solutions are defined as follows.

**Definition 13.2.5.** A function $\alpha \in X_\phi \cap C^2(\mathbb{R}^+)$ such that $\phi(\alpha') \in AC(\mathbb{R}_0^+)$ is said to be a lower solution of problem (13.1.1),(13.1.2) if

$$(\phi(\alpha'))'(t) + q(t)f(t, \alpha(t), \alpha'(t)) \geq 0$$

and

$$L(\alpha, \alpha(0), \alpha'(0)) \geq 0, \quad \alpha'(+\infty) < B, \qquad (13.2.9)$$

where $B \in \mathbb{R}$.

A function $\beta \in X_\phi \cap C^2(\mathbb{R}^+)$ is an upper solution if it satisfies the reversed inequalities.

The following condition is applied for well-ordered lower and upper solutions of problem (13.1.1),(13.1.2).

**(H5)** There are $\alpha$ and $\beta$ lower and upper solutions of (13.1.1),(13.1.2), respectively, such that

$$\alpha(t) \leq \beta(t), \quad \forall t \geq 0. \qquad (13.2.10)$$

Throughout the proof of the main result, a modified and perturbed problem will be considered. It is given by

$$\begin{cases} (\varphi(u'(t)))' + q(t)f(t, \delta_0(t, u), \delta_1(t, u')) = 0, \\ u(0) = \delta_0(0, u(0) + L(u, u(0), u'(0))), \\ u'(+\infty) = B \end{cases} \quad (13.2.11)$$

with the truncature $\delta_0 : \mathbb{R}_0^+ \times \mathbb{R} \to \mathbb{R}$ is given by

$$\delta_0(t, y) = \begin{cases} \beta(t), & y > \beta(t), \\ y, & \alpha(t) \le y \le \beta(t), \\ \alpha(t), & y < \alpha(t), \end{cases} \quad (13.2.12)$$

and $\delta_1 : \mathbb{R} \to \mathbb{R}$ by

$$\delta_1(w) = \begin{cases} N, & w > N, \\ w, & -N \le w \le N, \\ -N, & w < -N, \end{cases} \quad (13.2.13)$$

where $N$ is defined in Lemma 13.2.2, for functions $f$ satisfying Nagumo's condition.

Consider $\varphi : \mathbb{R} \to \mathbb{R}$ given by (13.2.6) where $R := \max\{N, \|\alpha'\|_1, \|\beta'\|_1\}$, with $N$ given by (9.2.15).

The operator $T : X_\phi \to X_\phi$ associated to (13.2.11) can then be defined as

$$(Tu)(t) := \delta_0(0, u(0) + L(u, u(0), u'(0)))$$
$$+ \int_0^t \varphi^{-1} \left( \varphi(B) + \int_s^{+\infty} q(\tau)f(\tau, \delta_0(\tau, u), \delta_1(\tau, u'))d\tau \right) ds.$$

$$(13.2.14)$$

One of the essential steps is to prove that the operator $T$ has a fixed point. However, the function $q$ may, or may not, be singular at the origin. In this way two results are presented: one for the regular case, where $q$ is not singular when $t = 0$, and another result for the singular case.

First, let us start by presenting some lemmas for the regular case.

**Lemma 13.2.6. (Regular case).** *Assume that $q : \mathbb{R}_0^+ \to \mathbb{R}_0^+$ is continuous and that conditions* (H1)–(H3) *and* (H5) *hold. Then the operator $T$ is well defined.*

**Proof.** For any $u \in X_\phi$ there is $K > 0$ such that $\|u\|_{X_\phi} < K$.

From (13.2.11) and (13.2.12)

$$\lim_{t \to +\infty} \frac{(Tu)(t)}{e^{\theta t}} \le \lim_{t \to +\infty} \frac{\beta(0)}{e^{\theta t}}$$

$$+ \lim_{t \to +\infty} \frac{\int_0^t \varphi^{-1}(\varphi(B) + \int_s^{+\infty} q(\tau) f(\tau, \delta_0(\tau, u), \delta_1(\tau, u')) d\tau) ds}{e^{\theta t}}$$

$$\le \lim_{t \to +\infty} \frac{\int_0^t \varphi^{-1}(\varphi(B) + \int_s^{+\infty} q(\tau) f(\tau, \delta_0(\tau, u), \delta_1(\tau, u')) d\tau) ds}{e^{\theta t}}.$$

As $\delta_0(\tau, u)$ and $\delta_1(\tau, u')$ are bounded, by (H2), then

$$f(\tau, \delta_0(\tau, u), \delta_1(\tau, u'))$$

is uniformly bounded. Let us define

$$S_K := \sup_{t \ge 0} \left\{ f(t, x, y), t \ge 0, |x| \in (0, K_0), |y| \in (0, N) \right\}, \qquad (13.2.15)$$

with

$$K_0 = \max \left\{ \|\alpha\|_0, \|\beta\|_0 \right\} \qquad (13.2.16)$$

and $N$ given by (13.2.3).

Remark that $S_K$ does not depend on $u$.

From (H3), a real number $k_1$ can be defined such that

$$\int_s^{+\infty} q(\tau) S_K d\tau := k_1. \qquad (13.2.17)$$

As $\varphi$ is nondecreasing, the previous inequality now becomes

$$\lim_{t \to +\infty} \frac{(Tu)(t)}{e^{\theta t}} \le \lim_{t \to +\infty} \frac{\int_0^t \varphi^{-1}(\varphi(B) + S_K \int_s^{+\infty} q(\tau) d\tau) ds}{e^{\theta t}}$$

$$\le \lim_{t \to +\infty} \frac{\int_0^t \varphi^{-1}(\varphi(B) + k_1) ds}{e^{\theta t}}$$

$$\le \lim_{t \to +\infty} \frac{\varphi^{-1}(\varphi(B) + k_1) t}{e^{\theta t}} = 0. \qquad (13.2.18)$$

For

$$\lim_{t \to +\infty} (Tu)'(t) = \varphi^{-1} \left( \varphi(B) + \int_t^{+\infty} q(\tau) f(\tau, \delta_0(\tau, u), \delta_1(\tau, u')) d\tau \right)$$

$$= B < +\infty.$$

Therefore, $T$ is well defined. $\qquad \square$

**Lemma 13.2.7. (Regular case).** *Assume that* $q : \mathbb{R}_0^+ \to \mathbb{R}_0^+$ *is continuous and that conditions* (H1)–(H5) *hold. Then the operator* $T$ *is continuous.*

**Proof.** Consider a convergent sequence $u_n \to u \in X_\phi$.

By the arguments used in the previous lemma, the upper bounds are uniform and, therefore, do not depend on $n$.

Defining

$$\Theta := \varphi(B) + \int_s^{+\infty} q(\tau) f(\tau, \delta_0(\tau, u_n), \delta_1(\tau, u_n')) d\tau$$

and as $\varphi$ is continuous, by (H2) and Lebesgue's Dominated Convergence Theorem, one has

$$\|(Tu_n) - (Tu)\|_0$$

$$= \sup_{t \geq 0} e^{-\theta t} \left| \begin{array}{l} \delta\left(0, u_n\left(0\right) + L\left(u_n, u_n\left(0\right), u_n'\left(0\right)\right)\right) + \int_0^t \varphi^{-1}(\Theta) ds \\ -\delta\left(0, u\left(0\right) + L\left(u, u\left(0\right), u'\left(0\right)\right)\right) - \int_0^t \varphi^{-1}(\Theta) ds \end{array} \right| \to 0,$$

as $n \to +\infty$, and

$$\|(Tu_n)' - (Tu)'\|_1$$

$$\leq \sup_{t \geq 0} \left| \begin{array}{l} \varphi^{-1}(\varphi(B) + \int_t^{+\infty} q(\tau) f(\tau, \delta_0(\tau, u_n), \delta_1(\tau, u_n')) d\tau) \\ -\varphi^{-1}(\varphi(B) + \int_t^{+\infty} q(\tau) f(\tau, \delta_0(\tau, u), \delta_1(\tau, u')) d\tau) \end{array} \right| \to 0,$$

as $n \to +\infty$.

Therefore, $T$ is continuous in $X_\phi$. $\qquad\square$

**Lemma 13.2.8.** *The operator* $T$ *is compact.*

**Proof.** The idea in this proof is to apply Lemma 13.2.4. For that, it is important to show that the operator $T$ is equicontinuous and equiconvergent at $+\infty$.

Let us consider $t_1, t_2 \in (0, T_0)$, where $T_0 > 0$ and $t_1 < t_2$.

Defining $\Theta := \varphi(B) + \int_s^{+\infty} q(\tau) f(\tau, \delta_0(\tau, u), \delta_1(\tau, u')) d\tau$, then, for $\theta > 0$,

$$\left| \frac{(Tu)(t_1)}{e^{\theta t_1}} - \frac{(Tu)(t_2)}{e^{\theta t_2}} \right| \leq \max\{|\alpha(0)|, |\beta(0)|\} \frac{e^{\theta t_2} - e^{\theta t_1}}{e^{\theta(t_1+t_2)}}$$

$$+ \left| \frac{e^{\theta t_2} - e^{\theta t_1}}{e^{\theta(t_1+t_2)}} \int_0^{t_1} \varphi^{-1}(\Theta) ds \right| + \left| \frac{e^{\theta t_1} \int_{t_1}^{t_2} \varphi^{-1}(\Theta) ds}{e^{\theta(t_1+t_2)}} \right|$$

$$\leq \max\{|\alpha(0)|, |\beta(0)|\} \frac{e^{\theta t_2} - e^{\theta t_1}}{e^{\theta(t_1+t_2)}}$$

$$+ \left| \frac{e^{\theta t_2} - e^{\theta t_1} \int_0^{t_1} \varphi^{-1}(\varphi(B) + S_K \int_s^{+\infty} q(\tau) d\tau)}{e^{\theta(t_1+t_2)}} \right|$$

$$+ \left| \frac{e^{\theta t_1} \int_{t_1}^{t_2} \varphi^{-1}(\varphi(B) + S_K \int_s^{+\infty} q(\tau) d\tau)}{e^{\theta(t_1+t_2)}} \right| \to 0,$$

as $t_1 \to t_2$.

Also, as $\varphi^{-1}$ is continuous, defining $F := q(\tau) f(\tau, \delta_0(\tau, u), \delta_1(\tau, u'))$, by (13.2.15) and (13.2.17),

$$|(Tu)'(t_1) - (Tu)'(t_2)| = \left| \varphi^{-1}\left( \int_{t_1}^{+\infty} F d\tau \right) - \varphi^{-1}\left( \int_{t_2}^{+\infty} F d\tau \right) \right| \to 0,$$

as $t_1 \to t_2$. Therefore, $T$ is equicontinuous.

For the equiconvergence at $+\infty$ of the operator $T$, one has, by (13.2.18),

$$\left| \frac{(Tu)(t)}{e^{\theta t}} - \lim_{t \to +\infty} \frac{(Tu)(t)}{e^{\theta t}} \right| = \left| e^{-\theta t} \int_0^t \varphi^{-1}(\Theta) ds \right| \to 0,$$

as $t \to +\infty$. For

$$\left| (Tu)'(t) - \lim_{t \to +\infty} (Tu)'(t) \right| = \left| \varphi^{-1}(\Theta) - \lim_{t \to +\infty} \varphi^{-1}(\Theta) \right|$$

it tends to 0 as $t \to +\infty$, from (H3) and the continuity of $\varphi^{-1}$.

As $T$ is equicontinuous and equiconvergent, then from Lemma 13.2.4, $T$ is compact. □

Now let us consider the singular case.

**Lemma 13.2.9. (Singular case).** *Let $q$ be singular at $t = 0$. Then the operator $T$ given by (13.2.14) is completely continuous.*

**Proof.** For each $n \geq 1$ and $\Theta := \varphi(B) + \int_s^{+\infty} q(\tau)f(\tau, \delta_0(\tau, u), \delta_1(\tau, u'))d\tau$ let us define the approximating operator $T_n : X_\phi \to X_\phi$ given by

$$(T_n u)(t) := \delta_0(0, u(0) + L(u, u(0), u'(0))) + \int_{\frac{1}{n}}^t \varphi^{-1}(\Theta)ds. \qquad (13.2.19)$$

In this case, it is sufficient to show that $T_n$ tends to $T$ on $X_\phi$. In fact, from (H1)–(H3), (13.2.15) and (13.2.17), one has

$$\left| \frac{(Tu)(t)}{e^{\theta t}} - \frac{(T_n u)(t)}{e^{\theta t}} \right| = \left| \frac{\int_0^{\frac{1}{n}} \varphi^{-1}(\Theta)ds}{e^{\theta t}} \right|$$

$$\leq \frac{\int_0^{\frac{1}{n}} \varphi^{-1}(\varphi(B) + S_K \int_s^{+\infty} q(\tau)d\tau)}{e^{\theta t}} \to 0,$$

as $n \to +\infty$, and

$$|(Tu)'(t) - (T_n u)'(t)|$$

$$= \left| \begin{array}{l} \varphi^{-1}(\varphi(B) + \int_{\frac{1}{n}}^{+\infty} q(\tau)f(\tau, \delta_0(\tau, u), \delta_1(\tau, u'))d\tau) \\ -\varphi^{-1}(\varphi(B) + \int_0^{+\infty} q(\tau)f(\tau, \delta_0(\tau, u), \delta_1(\tau, u'))d\tau) \end{array} \right| \to 0,$$

as $n \to +\infty$.

Hence, the operator $T$ is completely continuous. $\qquad \square$

## 13.3. Existence and localization result

In this section, the existence and location result for (13.1.1), (13.1.2) is proved.

**Theorem 13.3.1.** *Let* $f : \mathbb{R}_0^+ \times \mathbb{R}^2 \to \mathbb{R}$ *and* $q : \mathbb{R}_0^+ \to \mathbb{R}$ *be both continuous functions, where* $q$ *can have a singularity when* $t = 0$, *and* $f$ *verifies the Nagumo conditions* (13.2.1) *and* (13.2.2). *If conditions* (H1)–(H5) *are satisfied, then problem* (13.1.1),(13.1.2) *has at least one solution* $u \in X_\phi$ *and there exists* $N > 0$ *such that*

$$\alpha(t) \leq u(t) \leq \beta(t) \quad and \quad -N < u'(t) < N, \ \forall t \geq 0.$$

**Proof.**

**Claim 1.** *Every solution* $u$ *of* (13.2.11) *verifies* $\alpha(t) \leq u(t) \leq \beta(t)$ *and there is* $N > 0$ *such that* $-N < u'(t) < N, \forall t \geq 0$.

Let $u \in X_\phi$ be a solution of the modified problem (13.2.11) and suppose, by contradiction, that there exists $t > 0$ such that $\alpha(t) > u(t)$. Therefore,

$$\inf_{t \geq 0}(u(t) - \alpha(t)) < 0.$$

Suppose that this infimum is attained as $t \to +\infty$. Therefore,

$$\lim_{t \to +\infty}(u'(t) - \alpha'(t)) = u'(+\infty) - \alpha'(+\infty) \leq 0.$$

By Definition 13.2.5, one gets the contradiction,

$$0 \geq u'(+\infty) - \alpha'(+\infty) = B - \alpha'(+\infty) > 0.$$

Analogously, the infimum does not happen at $t = 0$, otherwise the following contradiction holds:

$$0 > u(0) - \alpha(0) = \delta(0, u(0) + L(u, u(0), u'(0))) - \alpha(0) \geq 0.$$

Therefore, there are $t_* > 0$ and $t_0 < t_*$ such that

$$\min_{t \geq 0}(u(t) - \alpha(t)) := u(t_*) - \alpha(t_*) < 0,$$

$$u'(t_*) = \alpha'(t_*),$$

$$u(t) < \alpha(t), \quad u'(t) < \alpha'(t), \quad \forall t \in [t_0, t_*[,$$

and, by (H1),

$$\varphi(u'(t)) < \varphi(\alpha'(t)), \quad \forall t \in [t_0, t_*[. \tag{13.3.1}$$

So, for $t \in [t_0, t_*[$, by (13.2.11), (13.2.12), (13.2.6) and Definition 13.2.5, one has

$$(\varphi(u'(t)))' = -q(t)f(t, \delta_0(t, u), \delta_1(t, u'))$$
$$= -q(t)f(t, \alpha(t), \alpha'(t))$$
$$\leq (\phi(\alpha'(t)))' = (\varphi(\alpha'(t)))'.$$

Therefore, the function $\varphi(u'(t)) - \varphi(\alpha'(t))$ is nonincreasing on $[t_0, t_*[$ and

$$\varphi(u'(t_0)) - \varphi(\alpha'(t_0)) \geq \varphi(u'(t_*)) - \varphi(\alpha'(t_*)) = 0,$$

which is a contradiction with (13.3.1).

So, $u(t) \geq \alpha(t), \forall t \geq 0$.

Analogously, it can be shown that $u(t) \leq \beta(t), \forall t \geq 0$.

The first derivative inequalities are an immediate consequence of Lemma 13.2.2, taking

$$\gamma(t) = \frac{\alpha(t)}{e^{\theta t}} \quad \text{and} \quad \Gamma(t) = \frac{\beta(t)}{e^{\theta t}}, \quad \text{for } t \geq 0, \ \theta > 0.$$

From the lemmas in the previous section, one has that the operator $T$ is completely continuous, both for the singular and regular cases.

**Claim 2.** *The problem* (13.2.11) *has at least a solution* $u \in X_\phi$.

In order to apply Schauder's fixed-point theorem, we consider a closed and bounded set $D$ defined as

$$D = \{u \in X_\phi : \|u\|_X \leq \rho\},$$

with $\rho$ such that

$$\rho := \max\left\{K_0 + \sup_{t \in [0,+\infty)}\left(\frac{\varphi^{-1}\left(\varphi\left(B\right) + k_1\right)t}{e^{\theta t}}\right), \left|\varphi^{-1}\left(\varphi\left(B\right) + k_1\right)\right|\right\},$$

where $K_0$ is given by (13.2.16) and $k_1$ by (13.2.17).

For $u \in D$, arguing as in the proof of Lemma 13.2.6, as $\varphi^{-1}$ is increasing, we have, for $S_K$ given by (13.2.15),

$$\begin{aligned}
\|Tu\|_0 &= \sup_{t \in [0,+\infty)} \frac{|(Tu)(t)|}{e^{\theta t}} \\
&\leq \sup_{t \in [0,+\infty)}\left(K_0 + \frac{\int_0^t \varphi^{-1}\left(\varphi\left(B\right) + \int_s^\infty q\left(\tau\right)S_K\right)ds}{e^{\theta t}}\right) \\
&\leq \sup_{t \in [0,+\infty)}\left(K_0 + \frac{\int_0^t \varphi^{-1}\left(\varphi\left(B\right) + k_1\right)ds}{e^{\theta t}}\right) \\
&= \sup_{t \in [0,+\infty)}\left(K_0 + \frac{\varphi^{-1}\left(\varphi\left(B\right) + k_1\right)t}{e^{\theta t}}\right) \leq \rho,
\end{aligned}$$

and

$$\begin{aligned}
\left\|(Tu)'\right\|_1 &= \sup_{t \in [0,+\infty)} \left|(Tu)'(t)\right| \\
&\leq \sup_{t \in [0,+\infty)}\left|\begin{array}{l}\varphi^{-1}\left(\varphi\left(B\right)\right. \\ \left. + \int_0^\infty q\left(\tau\right)f\left(\tau, \delta_0(\tau, u), \delta_1(\tau, u')\right)d\tau\right)\end{array}\right| \\
&\leq \sup_{t \in [0,+\infty)}\left|\varphi^{-1}\left(\varphi\left(B\right) + k_1\right)\right| \leq \rho.
\end{aligned}$$

Therefore, $TD \subseteq D$.

Then by Schauder's fixed-point theorem (Theorem 1.2.6), $T$ has at least one fixed point $u \in X_\phi$, that is, the problem (13.2.11) has at least one solution $u \in X_\phi$.

**Claim 3.** *Every solution $u$ of the problem* (13.2.11) *is a solution of problem* (13.1.1),(13.1.2).

Let $u$ be a solution of the modified problem (13.2.11). By the last claim, function $u$ verifies equation (13.1.1).

Then, it is enough to prove the inequalities

$$\alpha(0) \leq u(0) + L(u, u(0), u'(0)) \leq \beta(0).$$

Suppose, by contradiction, that $\alpha(0) > u(0) + L(u, u(0), u'(0))$.
By (13.2.11) and (13.2.12),

$$u(0) = \delta_0(0, u(0) + L(u, u(0), u'(0))) = \alpha(0).$$

Therefore, by Claim 1, $u'(0) \geq \alpha'(0)$.
By (H4) and Definition 13.2.5, the following contradiction is obtained

$$0 > u(0) + L(u, u(0), u'(0)) - \alpha(0) \geq L(\alpha, \alpha(0), \alpha'(0)) \geq 0.$$

In a similar way one can prove that $u(0) + L(u, u(0), u'(0)) \leq \beta(0)$. $\square$

**Remark 13.3.2.** Theorem 13.3.1 still remains true for singular $\phi$-Laplacian equations. Indeed, from Nagumo condition and Lemma 13.2.2, for every $u$ solution of problem (13.2.11), $\|u'(t)\|_1 < N$, and, therefore, considering in (13.2.6), $R > N$, one has

$$\phi :] - N, N[ \rightarrow \mathbb{R} \quad \text{and} \quad \phi(u'(t)) \equiv \varphi(u'(t)), \ \forall t \in \mathbb{R}_0^+.$$

The control on the first derivative given by Nagumo condition and Lemma 13.2.2, which implies a subquadratic growth on the nonlinearity, can be overcome assuming stronger conditions on lower and upper solutions, as in the next theorem.

**Theorem 13.3.3.** *Let $f : \mathbb{R}_0^+ \times \mathbb{R}^2 \rightarrow \mathbb{R}$ and $q : \mathbb{R}_0^+ \rightarrow \mathbb{R}$ be both continuous functions, where $q$ can have a singularity when $t = 0$. Assume that there are $\alpha$ and $\beta$ lower and upper solutions of (13.1.1),(13.1.2), respectively, such that*

$$\alpha'(t) \leq \beta'(t), \quad \forall t \geq 0, \tag{13.3.2}$$

*and*

$$\alpha(0) \leq \beta(0). \tag{13.3.3}$$

*If conditions* (H1)–(H4) *are satisfied and*

$$f(t, \alpha(t), y) \leq f(t, x, y) \leq f(t, \beta(t), y), \qquad (13.3.4)$$

*for* $\alpha(t) \leq x \leq \beta(t)$ *and* $y \in \mathbb{R}$ *fixed, then problem* (13.1.1),(13.1.2) *has at least a solution* $u \in X_\phi$ *such that*

$$\alpha'(t) \leq u'(t) \leq \beta'(t), \quad \forall t \geq 0.$$

**Remark 13.3.4.** Condition (13.3.2) together with (13.3.3) imply (H5).

**Proof.** The proof follows analogous steps as in Claims 1 and 2 of Theorem 13.3.1, with $\varphi$ defined by

$$R := \max \{\|\alpha'\|_1, \|\beta'\|_1\}. \qquad (13.3.5)$$

It remains to prove that $\alpha'(t) \leq u'(t) \leq \beta'(t)$, $\forall t \geq 0$.

Assume that there is a $t \geq 0$ such that $u'(t) < \alpha'(t)$, and define $t_0 \geq 0$ as

$$\inf_{t \geq 0} (u'(t) - \alpha'(t)) := u'(t_0) - \alpha'(t_0) < 0. \qquad (13.3.6)$$

By (13.1.2), there is $t_1 \in (t_0, +\infty)$ such that $u'(t_1) = \alpha'(t_1)$.
By (13.3.4), for $t \in [t_0, t_1]$,

$$(\varphi(u'(t)))'(t) = -q(t) f(t, \delta_0(t, u), \delta_1(t, u')) = -q(t) f(t, \delta_0(t, u), \alpha'(t))$$
$$\leq -q(t) f(t, \alpha(t), \alpha'(t)) \leq (\phi(\alpha'(t)))' = (\varphi(\alpha'(t)))'.$$

Therefore, $\varphi(u'(t)) - \varphi(\alpha'(t))$ is nonincreasing on $[t_0, t_1]$ and

$$\varphi(u'(t_0)) - \varphi(\alpha'(t_0)) \geq \varphi(u'(t_1)) - \varphi(\alpha'(t_1)) = 0.$$

So, $\varphi(u'(t_0)) \geq \varphi(\alpha'(t_0))$, and by (H1), $u'(t_0) \geq \alpha'(t_0)$ which contradicts (13.3.6). That is, $\alpha'(t) \leq u'(t)$, $\forall t \geq 0$.

In the same way it can be shown that $u'(t) \leq \beta'(t)$, $\forall t \geq 0$. $\qquad \square$

**Remark 13.3.5.** Theorem 13.3.3 holds for singular $\phi$-Laplacian equations. If in (13.2.6). $R$ is considered given by (13.3.5), one has

$$\phi :] - R, R[ \to \mathbb{R} \quad \text{and} \quad \phi(u'(t)) \equiv \varphi(u'(t)), \quad \forall t \geq 0.$$

## 13.4.   Examples

In order to demonstrate the applicability of the results in this chapter two examples will follow. In the first one the nonlinearity $f$ satisfies the Nagumo conditions and, in the second one, this assumption is replaced by a monotone behavior in $f$.

In both cases, the null function is not a solution of the referred problem.

**Example A.** Consider for some $\theta > 0$ the nonlinear problem composed by the differential equation

$$\frac{u''(t)}{1+(u'(t))^2} - \frac{1}{1+t^2}\frac{u(t)(u'(t))^2}{1+u^2(t)} = 0, \quad t \geq 0, \tag{13.4.1}$$

and the functional boundary conditions

$$\max_{t\geq 0} \frac{|u(t)|}{e^{\theta t}} + (u'(0))^3 - u(0) = 0, \quad u'(+\infty) = \frac{1}{2}. \tag{13.4.2}$$

Remark that this problem (13.4.1),(13.4.2) is a particular case of (13.1.1)–(13.1.2) with

- $\phi(v) = \arctan v$;

- $f(t,x,y) = -\dfrac{xy^2}{1+x^2}$;

- $q(t) = \dfrac{1}{1+t^2}$;

- $L(u,x,y) = \max_{t\in\mathbb{R}_0^+} \frac{|u(t)|}{e^{\theta t}} + y^3 - x$;

- $B = \dfrac{1}{2}$.

We point out that:

- $f(t,x,y)$ and $q(t)$ verify (H2), (H3) and the Nagumo conditions (13.2.1) and (13.2.2) with $\psi(t) \equiv 1$ and $h(|y|) = y^2$;
- $L(u,x,y)$ satisfies (H4);
- the functions $\alpha(t) = 0,5$ and $\beta(t) = t+2$ are, respectively, lower and upper solutions of (13.4.1),(13.4.2) verifying (H5);
- as $\phi$ is a nonsurjective homeomorphism satisfying (H1), it can be extended by a surjective homeomorphism $\varphi$, like in (13.2.6), that is

$$\varphi(y) = \begin{cases} \arctan(y) & \text{if } |y| \leq R, \\ \dfrac{\arctan(R)}{R} y & \text{if } |y| > R, \end{cases}$$

with

$$R := \max \{\|\alpha'\|_1, \|\beta'\|_1\} = 1.$$

So, by Theorem 13.3.1, there is at least a solution $u$ of (13.4.1),(13.4.2) such that

$$0,5 \le u(t) \le t + 2, \quad \forall t \ge 0.$$

Moreover, this solution is unbounded and, from the location part, strictly positive in $\mathbb{R}_0^+$.

**Example B.** The functional problem

$$\begin{cases} 3(u'(t))^2 u''(t) + \dfrac{1}{1+t^3} \left( \arctan\left((u(t))^3\right) - 2\dfrac{(u'(t))^5}{1+|u'(t)|^5} \right) = 0, \quad t \ge 0, \\ \displaystyle\int_0^1 \dfrac{u(t)}{e^{\theta t}} dt - 5u(0) + u'(0) = 1, \\ u'(+\infty) = B, \end{cases}$$

for some $\theta > 0$ and $B > -1$, is a particular case of (13.1.1),(13.1.2) with

- $\phi(v) = v^3$;

- $f(t, x, y) = \arctan\left(x^3\right) - 2\dfrac{y^5}{1+|y|^5}$;

- $q(t) = \dfrac{1}{1+t^3}$;

- $L(u, x, y) = \displaystyle\int_0^1 \dfrac{u(t)}{e^{\theta t}} dt - 5x + y - 1.$

Remark that, in this case, $\phi$ is a surjective homeomorphism and $f$ does not satisfy the Nagumo conditions but it verifies (13.3.4).

As the functions $\alpha(t) = -t - 1$ and $\beta(t) \equiv 0$ are, respectively, lower and upper solutions of (13.4), satisfying assumptions (13.3.2) and (13.3.3), then, by Theorem 13.3.3, there is at least a solution $u$ of (13.4) such that

$$-t - 1 \le u(t) \le 0, \quad \forall t \ge 0.$$

Indeed, this solution is unbounded if $B \ne 0$ and bounded if $B = 0$, and, in any case, nonpositive in $\mathbb{R}_0^+$.

# Bibliography

[1] G. Adams and L. Lin, Beam on a tensionless elastic foundation, *J. Eng. Mech.* **113** (1986) 542–553.

[2] R. P. Agarwal and P. J. Wong, Lidstone polynomials and boundary value problems, *Comput. Math. Appl.* **17**(10) (1988) 1397–1421.

[3] R. P. Agarwal and D. O'Regan, *Infinite Interval Problems for Differential, Difference and Integral Equations* (Kluwer Academic Publishers, Glasgow, 2001).

[4] R. P. Agarwal and D. O'Regan, Non-linear boundary value problems on the semi-infinite interval: an upper and lower solution approach, *Mathematika* **49**(1–2) (2002) 129–140.

[5] R. P. Agarwal, O. Mustafa and Y. Rogovchenko, Existence and asymptotic behavior of solutions of a boundary value problem on an infinite interval, *Math. Comput. Modelling* **41** (2005) 135–157.

[6] R. P. Agarwal and P. J. Wong, Eigenvalues of complementary Lidstone boundary value problems, *Bound. Value Probl.* **2012**(49) (2012).

[7] R. P. Agarwal and P. J. Wong, Positive solutions of complementary Lidstone boundary value problems. *Electron. J. Qual. Theory Differ. Equ.* **60** (2012) 1–200.

[8] F. Alessio, P. Caldiroli, and P. Montecchiari, On the existence of Homoclinic orbits for the asymptotically periodic Duffing equation, *Topol. Methods Nonlinear Anal. J.* **12** (1998) 275–292.

[9] C. O. Alves, Existence of heteroclinic solutions for a class of non-autonomous second-order equation, *NoDEA Nonlinear Differential Equations Appl.* **2**(5) (2015) 1195–1212.

[10] C. Alves, P. Carrião, and L. Faria, Existence of homoclinic solutions for a class of second order ordinary differential equations, *Nonlinear Anal. Real World Appl.* **12** (2011) 2416–2428.

[11] C. J. Amick and J. F. Toland, Homoclinic orbits in the dynamic phase-space analogy of an elastic strut, *European J. Appl. Math.* **3** (1992) 97–114.

[12] D. Anderson and F. Minhós, A discrete fourth-order Lidstone problem with parameters, *Appl. Math. Comput.* **214** (2009) 523–533.

[13] J. Andres, G. Gabor, and L. Górniewicz Boundary value problems on infinite intervals, *Trans. Amer. Math. Soc.* **351**(12) (1999) 4861–4903.

[14] M. Arias, J. Campos, and C. Marcelli, Fastness and continuous dependence in front propagation in Fisher-KPP equations, *Discrete Cont. in. Dyn. Syst. Ser. B* **11**(1) (2009) 11–30.

[15] D. Aronson and H. Weinberger, Multidimensional nonlinear diffusion arising in population genetics, *Adv. Math.* **30** (1978) 33–76.

[16] C. Avramescu, Sur l'existence des solutions convergentes systèmes d'équations différentielles non linéaires, *Ann. Mat. Pura Appl.* **81**(1) (1969) 147–168.

[17] C. Avramescu and C. Vladimirescu, Existence of solutions to second order ordinary differential equations having finite limits at $\pm\infty$, *Electron. Differential Equations* **18** (2004) 1–12.

[18] W. Aziz, H. Leiva, and N. Merentes, Solutions of Hammerstein equations in the space $BV(I_a^b)$, *Quaestiones Math.* **37**(3) (2014) 359–370.

[19] C. Bai and C. Li, Unbounded upper and lower solution method for third-order boundary-value problems on the half-line, *Electron. J. Differential Equations* **119** (2009) 1–12.

[20] R. Bellman, *Mathematical Methods in Medicine* (World Scientific, Singapore, 1983).

[21] Z. Benbouziane, A. Boucherif, and S. Bouguima, Existence result for impulsive third order periodic boundary value problems, *Appl. Math. Comput.* **206** (2008) 728–737.

[22] A. Benmezai, J. Graef, and L. Kong, Positive solutions for abstract Hammerstein equations and applications, *Commun. Math. Anal.* **16**(1) (2014) 47–65.

[23] C. Bereanu and J. Mawhin, Boundary-value problems with non-surjective Φ-Laplacian and oneside bounded nonlinearity, *Adv. Differential Equations* **11** (2006) 35–60.

[24] C. Bereanu and J. Mawhin, Existence and multiplicity results for some nonlinear problems with singular $\phi$-laplacian, *J. Differential Equations*, **243** (2007) 536–557.

[25] B. Bianconi and F. Papalini, Non-autonomous boundary value problems on the real line. *Discrete Contin. Dyn. Syst.* **15** (2006) 759–776.

[26] A. Blackmore and G. Hunt, Principle of localized buckling for a strut on an elastoplastic foundation, *J. Appl. Mech.* **63** (1996) 234–239.

[27] D. Bonheure, Multitransition kinks and pulses for fourth order equations with a bistable nonlinearity, *Ann. Inst. H. Poincaré Anal. Non Linéaire* **21** (2004) 319–340.

[28] D. Bonheure and P. Torres, Bounded and homoclinic-like solutions of a second order singular differential equation, *Bull. London Math. Soc.* **44** (2012) 47–54.

[29] A. Boucherif, Positive solutions of second order differential equations with integral boundary conditions, *Discrete Contin. Dyn. Syst.* **2007** (2007) 155–159.

[30] A. Boucherif, Second order boundary value problems with integral boundary conditions, *Nonlinear Anal.* **70**(1) (2009) 364–371.

[31] A. Cabada and S. Heikkilä, Existence of solutions of third-order functional problems with nonlinear boundary conditions, *ANZIAM J.* **46**(1) (2004) 33–44.

[32] A. Cabada and F. Minhós and A. I. Santos, Solvability for a third order discontinuous fully equation with functional boundary conditions, *J. Math. Anal. Appl.* **322** (2006) 735–748.

[33] A. Cabada and J. Tomecek, Nonlinear second-order equations with functional implicit impulses and nonlinear functional boundary conditions, *J. Math. Anal. Appl.* **328** (2007) 1013–1025.

[34] A. Cabada and J. Tomeček, Extremal solutions for nonlinear functional $\phi$-Laplacian impulsive equations. *Nonlinear Anal.* **67** (2007) 827–841.

[35] A. Cabada and F. Minhós, Fully nonlinear fourth-order equations with functional boundary conditions, *J. Math. Anal. Appl.* **340** (2008) 239–251.

[36] A. Cabada, R. Pouso and F. Minhós, Extremal solutions to fourth-order functional boundary value problems including multipoint condition, *Nonlinear Anal. Real World Appl.* **10** (2009) 2157–2170.

[37] A. Cabada and J. Ángel Cid, Heteroclinic solutions for non-autonomous boundary value problems with singular $\Phi$-Laplacian operators, *Discrete Contin. Dyn. Syst. Suppl.* **2009** (2009) 118–122.

[38] A. Cabada, J. Fialho and F. Minhós, Non ordered lower and upper solutions to fourth order functional BVP, *Discrete Contin. Dyn. Syst. Suppl.* **2011** (Issue Special) (2011) 209–218.

[39] A. Cabada, J. Ángel Cid and B. M. Villamarín, Computation of Green's functions for Boundary Value Problems with Mathematica, *Appl. Math. Comput.* **219** (2012) 1919–1936.

[40] A. Cabada and G. Figueiredo, A generalization of an extensible beam equation with critical growth in $\mathbb{R}^n$, *Nonlinear Anal. - Real World Appl.* **20** (2014) 134–142.

[41] A. Cabada, G. Infante and F. A. Tojo, *Nonzero solutions of perturbed Hammerstein integral equations with deviated arguments and applications, Topol. Methods Nonlinear Anal.* (2016); DOI: 10.12775/TMNA.2016.005.

[42] A. Cabada and R. L. Pouso, Existence results for the problem $(\phi(u'))' = f(t, u, u')$ with nonlinear boundary conditions, *Nonlinear Anal.* **35** (1999) 221–231.

[43] A. Calamai, Heteroclinic solutions of boundary value problems on the real line involving singular $\Phi$-Laplacian operators, *J. Math. Anal. Appl.* **378** (2011) 667–679.

[44] A. Canada, P. Drabek and A. Fonda, *Handbook of Differential Equations: Ordinary Differential Equations*, Vol. 3, (Elsevier B.V., 2006).

[45] H. Carrasco and F. Minhós, Unbounded solutions for functional problems on the half-line, *Abstr. Appl. Anal.* **2016** (2016) Article ID 8987374, 7 pp.

[46] H. Carrasco and F. Minhós, On infinite elastic beam equations with unbounded nonlinearities, to appear.

[47] H. Carrasco and F. Minhós, Homoclinic solutions for nonlinear general fourth order differential equations, to appear.

[48] A. Champneys and G. Lord, Computation of homoclinic solutions to periodic orbits in a reduced water-wave problem, *Physica D.* **102** (1997) 101–124.

[49] C. E. Chidume, C. O. Chidume and M. S. Minjibir, A new method for proving existence theorems for abstract Hammerstein equations, *Abstr. Appl. Anal.* **2015** (2015), Article ID 627260, 7 pp.

[50] S. W. Choi, T. S. Jang, Existence and uniqueness of nonlinear deflections of an infinite beam resting on a non-uniform nonlinear elastic foundation, *Bound. Value Probl.* **2012**(5) (2012) 24 pp.

[51] C. Corduneanu, Integral Equations and Stability of Feedback Systems (Academic Press, New York, 1973).

[52] P. Coullet, C. Elphick and D. Repaux, Nature of spatial chaos, *Phys. Rev. Lett.* **58** (1987) 431–434.

[53] G. Cupini, C. Marcelli and F. Papalini, On the solvability of a boundary value problem on the real line, *Bound. Value Probl.* **2011** (2011) 26.

[54] G. Dajun, Multiple positive solutions for first order impulsive singular integro-differential equations on the half-line, *Acta Math. Sci.* **32B**(6) (2012) 2176–2190.

[55] Y. Ding and C. Lee, Homoclinics for asymptotically quadratic and superquadratic Hamiltonian systems, *Nonlinear Anal.* **71** (2009) 1395–1413.

[56] W. Ding, J. Mi, and M. Han, Periodic boundary value problems for the first order impulsive functional differential equations, *Appl. Math. Comput.* **165** (2005) 433–446.

[57] S. Djebali and K. Mebaraki, Existence and multiplicity results for singular $\phi$-Laplacian BVP's on the positive half-line, *Electron. J. Differential Equations*, **2009** (2009) 1–13.

[58] L. Erbe, Q. Kong and B. Zhang, Oscillation Theory for Functional Differential Equations (Marcel Dekker, Inc., 1995).

[59] C. Fabry and R. Manásevich, Equations with a $p$-Laplacian and an asymmetric nonlinear term, *Discrete Contin. Dynam. Syst.* **7** (2001) 545–557.

[60] H. Feng, D. Ji and W. Ge, Existence and uniqueness of solutions for a fourth-order boundary value problem, *Nonlinear Anal.* **70** (2009) 3761–3566.

[61] L. Ferracuti and F. Papalini, Boundary value problems for strongly nonlinear multivalued equations involving different $\Phi$-Laplacians, *Adv. Differ. Equ.* **14** (2009) 541–566.

[62] J. Fialho and F. Minhós, Existence and location results for hinged beams with unbounded nonlinearities, *Nonlinear Anal.* **71** (2009) 1519–1525.

[63] J. Fialho and F. Minhós, On higher order fully periodic boundary value problems, *J. Math. Anal. Appl.* **395** (2012) 616–625.

[64] J. Fialho and F. Minhós, The role of lower and upper solutions in the generalization of Lidstone problems, *Discrete Contin. Dyn. Syst. Suppl.* **2013** (2013) 217–226.

[65] J. Fialho and F. Minhós, Higher order functional boundary value problems without monotone assumptions, *Bound. Value Probl.* **2013**(81) (2013) 81.

[66] J. Fialho and F. Minhós, *High Order Boundary Value Problems: Existence, Localization and Multiplicity Results*, Mathematics Research Developments Series (Nova Science Publishers, Inc., New York, 2014).

[67] J. Fialho and F. Minhós, Fourth order impulsive periodic boundary value problems, *Differential Equations Dynam Systems*, **23**(2) (2015) 117–127.

[68] J. Fialho, F. Minhós and H. Carrasco Singular and classical second order $\phi$-Laplacian equations on the half-line with functional boundary conditions, to appear.

[69] D. Fu and W. Ding, Existence of positive solutions of third-order boundary value problems with integral boundary conditions in Banach spaces, *Adv. Differ.* **2013**(65) (2013).

[70] B. H. Gilding and R. Kersner, *Travelling Waves in Nonlinear Diffusion–Convection–Reaction* (Birkhäuser, Basel, 2004).

[71] J. Graef, L. Kong and F. Minhós, Higher order boundary value problems with $\phi$-Laplacian and functional boundary conditions, *Comput. Math. Appl.* **61** (2011) 236–249.

[72] J. Graef, L. Kong, F. Minhós and J. Fialho, On the lower and upper solution method for higher order functional boundary value problems, *Appl. Anal. Discrete Math.* **5**(1) (2011) 133–146.

[73] M. Gregus, *Third Order Linear Differential Equations*, Mathematics and its Applications (Reidel Publishing Co., Dordrecht, 1987).

[74] M. R. Grossinho, F. Minhós and S. Tersian, Positive homoclinic solutions for a class of second order differential equations, *J. Math. Anal. Appl.* **240** (1999) 163–173.

[75] M. R. Grossinho, F. Minhós and A. I. Santos, A note on a class of problems for a higher order fully nonlinear equation under one sided Nagumo type condition, *Nonlinear Anal.* **70** (2009) 4027–4038.

[76] J. Guckenheimer and P. Holmes, *Nonlinear Oscillations, Dynamical Systems, and Bifurcations of Vector Fields* (Springer, New York, 1983).

[77] C. P. Gupta, Existence and uniqueness theorems for the bending of an elastic beam equation, *Appl. Anal.* **26** (1988) 289–304.

[78] P. Habets and F. Zanolin, Upper and lower solutions for a generalized Emden–Fowler equation, *J. Math. Anal. Appl.* **181** (1994) 684–700.

[79] J. Hale, *Theory of Functional Differential Equations*, Applied Mathematical Sciences, Vol. 3 (Springer, 1977).

[80] A. Hammerstein, Nichtlineare Integralgleichungen nebst Anwendungen, *Acta Math.* **54** (1929) 117–176.

[81]  J. Han, Y. Liu and J. Zhao, Integral boundary value problems for first order nonlinear impulsive functional integro-differential differential equations, *Appli. Math. Comput.* **218** (2012) 5002–5009.

[82]  C. Harley and E. Momoniat, First integrals and bifurcations of a Lane–Emden equation of the second kind, *J. Math. Anal. Appl.* **344** (2008) 757–764.

[83]  G. W. Hunt, H. M. Bolt, and J. M. T. Thompson, Localization and the dynamical phase-space analogy, *Proc. R. Soc. Lond. Ser. A Math. Phys. Eng. Sci.* **425** (1989) 245–267.

[84]  G. Infante and P. Pietramala, Existence and multiplicity of non-negative solutions for systems of perturbed Hammerstein integral equations, *Nonlinear Anal.* **71** (2009) 1301–1310.

[85]  M. Izydorek and J. Janczewska, Homoclinic solutions for a class of the second order Hamiltonian systems, *J. Differential Equations* **219** (2005) 375–389.

[86]  T. S. Jang, H. S. Baek and J. K. Paik, A new method for the non-linear deflection analysis of an infinite beam resting on a non-linear elastic foundation, *Internat. J. Non-linear Mech.* **46** (2011) 339–346.

[87]  T. S. Jang, A new semi-analytical approach to large deflections of Bernoulli–Euler–v. Karman beams on a linear elastic foundation: Nonlinear analysis of infinite beams, *Internat. J. Mech. Sci.* **66** (2013) 22–32.

[88]  J. Jiang, L. Liu and Y. Wu, Second-order nonlinear singular Sturm Liouville problems with integral boundary conditions, *Appl. Math. Comput.* **215** (2009) 1573–1582.

[89]  J. Sun and T.-F. Wu, Two homoclinic solutions for a nonperiodic fourth order differential equation with a perturbation, *J. Math. Anal. Appl.* **413** (2014) 622–632.

[90]  K. A. Khachatryan and T. E. Terdzhyan, On the solvability of one class of nonlinear integral equations in $L_1(0, +\infty)$, *Siberian Adv. Math.* **25**(4) (2015) 268–275.

[91]  V. Kolmanovskii, A. Myshkis, *Applied Theory of Functional Differential Equations*, Mathematics and its Applications, Vol. 85 (Springer, 1992).

[92]  A. Kolmogorov, I. Petrovsky and N. Piskunov, Étude de l'équation de la diffusion avec croissance de la quantité de la matière et son application à un problème biologique, *Moscow Univ. Bull. Math.* **1** (1937) 1–25.

[93]  L. Kong and J. Wong, Positive solutions for higher order multi-point boundary value problems with nonhomogeneous boundary conditions, *J. Math. Anal. Appl.* **367** (2010) 588–611.

[94]  K. Q. Lan and W. Lin, Positive solutions of systems of singular Hammerstein integral equations with applications to semilinear elliptic equations in annuli, *Nonlinear Anal.* **74** (2011) 7184–7197.

[95]  V. Lakshmikantham, D. Baĭnov and P. Simeonov, Theory of impulsive differential equations, Series in Modern Applied Mathematics Vol. 6 (World Scientific Publishing Co., Inc., 1989).

[96]  J. Lega, J. Moloney and A. Newell, Swift–Hohenberg for lasers, *Phys. Rev. Lett.* **73** (1994) 2978–2981.

[97] C. Li, Remarks on homoclinic solutions for semilinear fourth-order ordinary differential equations without periodicity, *Appl. Math. J. Chinese Univ.* **24** (2009) 49–55.

[98] H. Lian, P. Wang and W. Ge, Unbounded upper and lower solutions method for Sturm-Liouville boundary value problem on infinite intervals, *Nonlinear Anal.* **70** (2009) 2627–2633.

[99] H. Lian and J. Zhao, Existence of unbounded solutions for a third-order boundary value problem on infinite intervals, *Discrete Dynam. Nature Soc.* **2012** (2012), Article ID 357697, 14 pp.

[100] H. Lian, J. Zhao and R. P Agarwal, Upper and lower solution method for nth-order BVPs on an infinite interval, *Bound. Value Probl.* **2014** (2014)100, 17 pp.

[101] Y. Liu, Existence of solutions of boundary value problems for coupled singular differential equations on whole lines with impulses, *Mediterr. J. Math.* **12** (2015) 697–716.

[102] Y. Liu, Solvability of boundary value problems for singular quasi-Laplacian differential equations on the whole line, *Math. Models Anal.* **17** (2012) 423–446.

[103] Y. Liu and S. Chen, Existence of bounded solutions of integral boundary value problems for singular differential equations on whole lines, *Internat. J. Math.* **25** (8) (2014) 1450078, 28 pp.

[104] Y. Liu and W. Ge, Solutions of a generalized multi-point conjugate BVPs for higher order impulsive differential equations. *Dynam. Syst. Appl.* **14** (2005) 265–279.

[105] L. Liu, Z. Wang and Y. Wu, Multiple positive solutions of the singular boundary value problems for second-order differential equations on the half-line, *Nonlinear Anal.* **71** (2009) 2564–2575.

[106] H. Lu, L. Sun and J. Sun, Existence of positive solutions to a non-positive elastic beam equation with both ends fixed, *Bound. Value Probl.* **2012**(56) (2012).

[107] P. Maheshwarin and S. Khatri, Nonlinear analysis of infinite beams on granular bed-stone column-reinforced earth beds under moving loads, *Soils Foundations* **52**(1) (2012) 114–125.

[108] P. K. Maini, L. Malaguti, C. Marcelli and S. Matucci, Diffusion-aggregation processes with mono-stable reaction terms, *Discrete Contin. Dyn. Syst. (B)*, **6**(5) (2006) 1175–1189.

[109] A. K. Mallik, S. Chandra, S. Sarvesh and B. Avinash, Steady-state response of an elastically supported infinite beam to a moving load, *J. Sound Vibration* **291** (2006) 1148–1169.

[110] C. Marcelli and F. Papalini, Heteroclinic connections for fully non-linear non-autonomous second-order differential equations, *J. Differential Equations* **241** (2007) 160–183.

[111] C. Marcelli, Existence of solutions to boundary value problems governed by general non-autonomous nonlinear differential operators, *Electron. J. Differential Equations* **2012**(171) (2012) 1–18.

[112] C. Marcelli, The role of boundary data on the solvability of some equations involving non-autonomous nonlinear differential, operators, *Bound. Value Probl.* **2013**(252) (2013).

[113] F. Minhós, Location results: An under used tool in higher order boundary value problems, *AIP Conf. Proc.* **1124** (2009) 244–253.

[114] F. Minhós, Existence of extremal solutions for some fourth order functional BVPs, *Commun. Appl. Anal.* **15** (2011) 547–556.

[115] F. Minhós, Sufficient conditions for the existence of heteroclinic solutions for $\varphi$-Laplacian differential equations, *Complex Variables Elliptic Equations* **2016**(62) (2017) 12 pp.

[116] F. Minhós and R. Carapinha, On higher order nonlinear impulsive boundary value problems, *Discrete Contin. Dyn. Syst. Suppl.* **2015** (2015) 851–860.

[117] F. Minhós and H. Carrasco, Solvability of higher-order BVPs in the half-line with unbounded nonlinearities, *Discrete Contin. Dyn. Syst. Suppl.* **2015** (2015) 841–850.

[118] F. Minhós and H. Carrasco, Existence of homoclinic solutions for nonlinear second-order problems, *Mediterr. J. Math.* **13**(6) (2016) 3849–3861.

[119] F. Minhós, T. Gyulov and A. I. Santos, Existence and location result for a fourth order boundary value problem, *Discrete Contin. Dyn. Syst. Suppl.* **2005** (2005) 662–671.

[120] F. Minhós, T. Gyulov and A. I. Santos Lower and upper solutions for a fully nonlinear beam equations, *Nonlinear Anal.* **71** (2009) 281–292.

[121] F. Minhós and J. Fialho, On the solvability of some fourth-order equations with functional boundary conditions, *Discrete Contin. Dyn. Syst. Suppl.* **2009** (2009) 564–573.

[122] Y. Momoya, S. Etsuo and F.Tatsuoka, Deformation characteristics of railway roadbed and subgrade under moving-wheel load, *Soils Foundations* **45**(4) (2005) 99–118.

[123] N. S. Papageorgiou and F. Papalini, Pairs of positive solutions for the periodic scalar $p$-Laplacian, *J. Fixed Point Theory* **5** (2009) 157–184.

[124] M. Pei, S. Chang and Y. S. Oh, Solvability of right focal boundary value problems with superlinear growth conditions, *Bound. Value Prob.* **2012** (2012) 60.

[125] M. A. Peletier, Sequential buckling: a variational analysis, *SIAM J. Math. Anal.* **32** (2001) 1142–1168.

[126] L. A. Peletier and W. C. Troy, *Spatial Patterns. Higher Order Models in Physics and Mechanics*, Progress in Nonlinear Differential Equations Applications, Vol. 45 (Birkhäuser Boston Inc., Boston, MA, 2001).

[127] M. del Pino, M. Elgueta and R. Manásevich, A homotopic deformation along p of a Leray-Schauder degree result and existence for $(|u'|^{p-2}u')' + f(t,u) = 0$, $u(0) = u(T) = 0$, $p > 1$, *J. Differential Equations* **80** (1989) 1–13.

[128] B. Przeradzki, The existence of bounded solutions for differential equations in Hilbert spaces, *Ann. Polon. Math.* **56**(2) (1992) 103–121.

[129] I. Rachůnková and M. Tvrdý, Existence results for impulsive second-order periodic problems. *Nonlinear Anal.* **59** (2004) 133–146.

[130] A. Salas and J. Castillo, Exact Solution to Duffing Equation and the Pendulum Equation, *Appl. Math. Sci.* **8**(76) (2014) 8781–8789.

[131] A. M. Samoilenko and N. A. Perestyuk,*Impulsive Differential Equations* (World Scientific, Singapore, 1995).

[132] S. Santra and J. C. Wei, Homoclinic solutions for fourth-order traveling wave equations, *SIAM J. Math. Anal.* **41** (2009) 2038–2056

[133] C. G. Small, *Functional Equations and How to Solve Them*, Problem Books in Mathematics (Springer, 2007).

[134] D. Smets and J. B. van den Berg, Homoclinic solutions for Swift–Hohenberg and suspension bridge type equations, *J. Differential Equations* **184** (2002) 78–96.

[135] J. Sun, H. Chen and J. J. Nieto, Homoclinic solutions for a class of subquadratic second-order Hamiltonian systems, *J. Math. Anal. Appl.* **373** (2011) 20–29.

[136] Y. Sun, Y. Sun and L. Debnath, On the existence of positive solutions for singular boundary value problems on the half line, *Appl. Math. Lett.* **22** (2009) 806–812.

[137] J. B. Swift and P. C. Hohenberg, Hydrodynamic fluctuations at the convective instability, *Phys. Rev. A* **15** (1977) 319–328.

[138] S. Tersian and J. Chaparova, Periodic and homoclinic solutions of extended Fisher–Kolmogorov equations, *J. Math. Anal. Appl.* **260** (2001) 490–506.

[139] E. O. Tuck and L. W. Schwartz, A boundary value problem from draining and coating flows involving a third-order differential equation relevant to draining and coating flows, *SIAM Rev.* **32** (1990) 453–469.

[140] J. Wang, J. Xu and F. Zhang, Homoclinic orbits for a class of Hamiltonian systems with superquadratic or asymptotically quadratic potentials, *Commun. Pure Appl. Anal.* **10** (2011) 269–286.

[141] M. X. Wang, A. Cabada and J. J. Nieto, Monotone method for nonlinear second order periodic boundary value problems with Carathéodory functions, *Ann. Polon. Math.* **58** (1993) 221–235.

[142] A.-W. Wazwaz, A domain decomposition method for a reliable treatment of the Emden–Fowler equation, *Appl. Math. Comput.* **161** (2005) 543–560.

[143] P. J. Y. Wong, Triple solutions of complementary Lidstone boundary value problems via fixed point theorems, *Bound. Value Probl.* **2014** (125) (2014).

[144] J. S. W. Wong, On the generalized Emden–Fowler equation, *SIAM Rev.* **17**(2) (1979) 339–360.

[145] L. Yan, J. Liu and Z. Luo, Existence and multiplicity of solutions for second-order impulsive differential equations on the half-line, *Adv. Differ. Equ.* **2013** (293) (2013).

[146] B. Yan, D. O'Regan and R.P. Agarwal, Unbounded solutions for singular boundary value problems on the semi-infinite interval: Upper and lower solutions and multiplicity, *J. Comput. Appl. Math.* **197** (2006) 365–386.

[147] B. Yan, D. O' Regan and R. P. Agarwal, Positive solutions for second order singular boundary value problems with derivative dependence on infinite intervals, *Acta Appl. Math.* **103** (2008) 19–57.

[148] Z. Yang, Positive solutions for a system of nonlinear Hammerstein integral equations and applications, *Appl. Math. Comput.* **218** (2012) 11138–11150.

[149] X. Yang and Y. Liu, Existence of unbounded solutions of boundary value problems for singular differential systems on whole line, *Bound. Value Problems* **2015**(42) (2015).

[150] Z. Yang and Z. Zhang, Positive solutions for a system of nonlinear singular Hammerstein integral equations via nonnegative matrices and applications, *Positivity* **16** (2012) 783–800.

[151] F. Yoruk and N. Aykut Hamal, Second-order boundary value problems with integral boundary conditions on the real line, *Electron. J. Differential Equations* **2014**(19) (2014) 1–13.

[152] E. Zeidler, *Nonlinear Functional Analysis and Its Applications, I: Fixed-Point Theorems* (Springer, New York, 1986).

[153] L. Zhang and W. Ge Solvability of a second order boundary value problem on an unbounded domain, *Appl. Math. E-Notes* **10** (2010) 40–46.

[154] Y. Zhang, Homoclinic solutions for a forced generalized Liénard system, *Adv. Differ. Equ.* **2012** (2012) 94.

[155] Z. Zhang and C. Zhang, Similarity solutions of a boundary layer problem with a negative parameter arising in steady two-dimensional flow for power-law fluids, *Nonlinear Anal.* **102** (2014) 1–13.

[156] C. Zhu and W. Zhang, Computation of bifurcation manifolds of linearly independent homoclinic orbits, *J. Differ. Equ.* **245** (2008) 1975–1994.

[157] S. Zhu, Q. Wu and X. Cheng, Numerical solution of the Falkner–Skan equation based on quasilinearization, *Appl. Math. Comput.* **215** (2009) 2472–2485.

# Index

**A**

Arzèla–Ascoli theorem, 136
asymptotic behavior, 44

**B**

Bernoulli–Euler–v. Karman problem,
  66, 131

**C**

cantilever beam, 53
Carathéodory sequences, 26
complementary Lidstone problems,
  72
completely continuous, 34

**E**

elastic beam, 5
Emden–Fowler equation, 149
equicontinuous, 16
equiconvergent at infinity, 28
Extended Fisher–Kolmogorov (EFK)
  equation, 58, 67

**F**

Falkner–Skan equation, 136
forced vibrations, 54

functional boundary conditions on
  the half-line, 153
functional boundary value problems,
  136

**G**

Green's functions, 9

**H**

half-line, 36
Hammerstein integral equations, 127
heteroclinic solutions, 41
homeomorphism, 83
homoclinic solutions, 42

**I**

impulsive problems, 25
infinite impulse moments, 25
infinite intervals, 5
infinite nonlinear beam, 67
integral equations of Hammerstein
  type, 83
integral operator, 129

**L**

$L^1$-Carathéodory function, 6
$p$-Laplacian, 83

Lebesgue's Dominated Convergence
  Theorem, 33
Lidstone problems, 72
lower solution, 10

**N**
Nagumo condition, 7
nonuniform elastic foundations, 67

**P**
$\phi$-Laplacian equations, 83
phase portrait, 65

**R**
railways, 78
relatively compact, 10

**S**
Schauder's fixed-point theorem, 10
sign-changing kernels, 127

singular $\phi$-Laplacian, 84
singular $\phi$-Laplacian equations, 125
subject, 10
Swift–Hohenberg (SH) equation, 58,
  67

**T**
truncature function, 52

**U**
uniformly bounded, 15
upper solution, 11

**Y**
Young's modulus, 131

Printed in the United States
By Bookmasters